KU-647-515

MEMBERS OF THE COMPUTER AND WORK DESIGN
RESEARCH UNIT – MANCHESTER BUSINESS SCHOOL

Past members

Roderick Atkin
Susan Gates
Valerie Kennedy-Browne
Andrew Pettigrew
Malcolm Rigby
Krystyna Weinstein

Present members

Pauline Harris
Paul Lloyd
Dorothy Mercer
Stephen Mills
Enid Mumford
Mary Weir

BUSINESS STRATEGY & PLANNING

Edited by Bernard Taylor

Implementing Strategic Decisions

Enid Mumford

Andrew Pettigrew

Longman
London and New York

LONGMAN GROUP LIMITED
London
*Associated companies, branches and representatives
throughout the world*

© Longman Group Limited

All rights reserved. No part of this publication
may be reproduced, stored in a retrieval system,
or transmitted in any form or by any means, electronic,
mechanical, photocopying, recording, or otherwise,
without the prior permission of the Copyright owner.

First published 1975

SBN 0 582 450063 (cased)

Set in 11 on 12 pt Imprint type,
and printed in Great Britain by
Western Printing Services Ltd,
Bristol

Contents

Part Three: Uncertainty and the innovator

Part Four: Uncertainty and the user

6 Innovator role definition and user uncertainty

7 Computer specialists and user departments – role definition and expectation

Preface

This book makes an original contribution to current thinking about business strategy and planning. Most writing about corporate strategy has concentrated on the problem of formulation. Enid Mumford and Andrew Pettigrew are concerned with the equally vital but neglected area of implementation.

The book is a research study into a fundamental management problem: the planning, organisation, and putting into operation of major, new projects. The subject of the investigation is an area of vital interest for public and private enterprises: the introduction of large-scale computer systems. The researchers have monitored the development of new computer systems over a period of two to five years from the decision to investigate the problem, through the search for a system, the decision to go ahead with a particular manufacturer, and the difficulties encountered in putting the system into operation.

The strategic decision

However, the theme of the book is not computers but planning, or more precisely the politics of planning. The writers set out to examine the management process surrounding the making of a major strategic decision. They examine the writings of economists, management scientists, psychologists and sociologists on the subject of strategic planning and decision-making. Then they investigate what happens in practice. The result is a fascinating book which throws light on an area of central interest to senior managers and administrators: the major capital investment decision. Four groups of fundamental questions are faced:

1 How are these big decisions made?
   This involves identifying the problem, collecting information, evaluating alternative solutions, as well as making the final decision.
2 What is the role of the expert in this process?
   Why do top management bring in experts in the first place?
   How are their roles defined?
   How do they achieve powerful positions and why do they sometimes lose their power?

3 What is the role of the operating managers and their staff in this situation? How do they relate to the experts?
What determines whether there is conflict or cooperation?

4 Why is it that some of these major projects take a long time to get 'off the ground', and others are quickly integrated into the normal operation?

Human aspects of planning

It will be clear that this is not a book on the financial or technical aspects of capital budgeting or project planning. The focus is on the human aspects of innovation, the personal and group relationships, and the internal politics. The authors challenge the conventional idea that planning is a 'nice tidy sequential process carried out logically and leading to the possibility of clear final choices'. They find that in practice there are elements of uncertainty which appear to make the application of coherent rational planning very difficult and very rare. However, top management and staff specialists tend frequently to ignore these areas where uncertainty is likely to arise – human needs, expectations and relationships – and treat planning as a technical problem.

Successful innovation

The conclusion is reached that the successful implementation of major strategic innovations such as the introduction of large-scale computer systems depends largely on management's handling of the human problems which arise.

1 There is usually a conflict of interests between top management, the staff specialists and various groups of operating management. The social system is not completely authoritarian and this results in a process of negotiation and bargaining. Mechanisms have to be created to facilitate and contain bargaining and negotiation between different groups within the planning procedures.

2 There is, of course, autonomous power and influence at lower levels in an organisation and groups which feel they are excluded from the decision process often find other ways of making their views known. Where serious attempts were made to involve interested parties at every level in the decisions that were being made, the protests were less serious.

3 The decision-making process can only be understood if it takes into account the personal values and interests of the individuals and groups concerned. It appears that decision-making is only effectively carried out if it provides some personal rewards for the decision maker. Planners must be aware of those solutions which will prove personally acceptable to the managers who have to operate the innovations.

4 Attempts at what is seen as rational decision-making frequently become distorted and overlaid by political issues which are not always recognised and made explicit. The planner should try to identify those who are par-

ticipating in the decision process and the different roles they are playing.

5 A danger associated with the use of computers has been an attraction to this technology as a means for change without significantly precise thought about the kinds of changes likely to prove most worth while. The result has been a proliferation of computer systems that have not been regarded as financially or administratively successful. The study emphasises the need for top management to set realistic goals and objectives to be achieved by the means of the innovation, and to recognise that these early goals may have to be changed as the situation changes and more information becomes available.

6 The search for alternative solutions to a problem often takes the form of a two-way, or 'mating' process: people with problems come together with people who have solutions. There is a need for top management to allocate resources for this search process, and to establish broad criteria by which alternative solutions may be compared.

7 For a strategic problem there is no correct or even best solution; in fact there is frequently no terminal state. This means that the choice between alternatives is usually the product of 'the strategic mobilisation of power resources', that is, of organisational politics. In practice, technical tests of feasibility tend to be fairly undemanding: for example, can we afford it? is there a likely financial benefit?

8 In the political process, the planner needs to identify at a fairly early stage the *human* consequences of alternative courses of action. All too often management concentrates on financial or technical matters and the human consequences to operational managers and their staff are largely ignored.

9 People operating as innovators suffer from stress and uncertainty because their organisation does not know how to use their skills. The specialist needs therefore to find out what assumptions others are making about his area of responsibility and establish how far these coincide with his own definition of his role. The more unstructured the situation, the more dependent he is on personal contacts and friendship.

10 The new specialist often acts as a catalyst by making managers aware of wider aspects of their jobs of which they had previously not been aware. This leads to stress and uncertainty for operating managers and there is also a need for agreement on the role the local manager should play during the planning and implementation period.

11 If the specialist does not try to satisfy the user's needs and objectives there will be conflict and resistance. If he tries to explain what he can offer in terms of the user department needs and objectives then cooperation may result.

Internal politics

Major strategic decisions tend to lead to a reallocation of resources, such as status, power and influence, in an enterprise, and they tend to become the focus of a good deal of political activity. A considerable amount of the book is devoted to the analysis of the internal politics associated with strategic choice.

Power is seen as deriving not only from hierarchical authority but also from other factors such as specialised knowledge, personal respect and acceptability, informal influence, external affiliations and historical relationships. Competition tends to centre round capital expenditure, the control of personnel and information, and new operations. Conflicts of interest as a result of differing group goals and power battles are seen as a normal part of organisational life, and the authors describe the various political strategies adopted by the protagonists, such as the creation of coalitions, withholding information, developing a rationale or case, discrediting the opposition, and preventing an opponent from retaliating.

Staff-line conflict

Another fascinating analysis concerns the conflict between the computer specialists and the operating management. These two groups are seen to have conflicts of interest, of personal values and of roles.

The specialist staff have a vested interest in innovation and change, no line authority, but a great deal of influence due to their technical knowledge and skill. The operating management are concerned with meeting production targets and maintaining day-to-day efficiency. They have power largely due to their position in the hierarchy and their control of resources.

The specialists tend to think of themselves as technical experts or scientific managers, and they judge their success in terms of financial or economic performance. The staff in operating units tend to be more concerned with human relationships and maintenance of their job satisfaction. They often seek a quiet life or a situation where they may undertake work that is familiar to them and in which they are proficient.

The definition of the roles and relationships between specialist staff and operating management, particularly in an innovative situation, tends to emerge over a period of time as a result of negotiation between the two groups. The innovator is seen as 'role-making' rather than 'role-taking'. His job is not tightly structured but depends on the individual, his success or failure in performance, top management's expectations, the expectations of other management groups, and external pressures on the enterprise.

Conclusion

What emerges from the book is a picture of the process of innovation in terms of human needs and interests, political struggles, and problems in defining roles and responsibilities in changing situations where there is a great deal at stake. This should be of immediate interest to all operating managers and staff specialists who are striving to introduce new operations in today's rapidly changing environment. One thinks immediately of specialists in corporate planning, organisation development, management services and of course computer services.

BERNARD TAYLOR
Administrative Staff College,
Henley.

Foreword

The research described in this book is concerned with analysing the way in which organisations cope with uncertainty. By uncertainty is meant situations which are not easily understood and predictable but complex and changing, and which contain elements likely to produce unexpected and unforeseeable consequences. Computer technology and the introduction of large-scale, advanced computer systems were chosen as the means for studying this subject because they were seen by the research team as developing very rapidly, as being very pervasive in the sense that most organisations would be using them for one purpose or another by the end of the 1970s, as having very considerable consequences for the structure and management of these organisations and for the people employed there, and, for all these reasons, as generating a great deal of turmoil and upheaval which the organisations would have to contain and cope with.

The research has taken five years to complete and has generously been financed throughout this period by the British Social Science Research Council to whom we should like to extend our grateful thanks. The individuals concerned with carrying out the research were Susan Gates, Valerie Kennedy-Browne, Dorothy Mercer, Enid Mumford, Andrew Pettigrew, Malcolm Rigby, Krystyna Weinstein and Mary Weir. The research was carried out in four industrial and commercial organisations which were introducing large-scale computer systems. The interest of these organisations in our research enabled us to observe the change processes in each situation over a number of years. We are most grateful to these organisations for tolerating research over such a long period of time and for their continual help with our investigation.

The plan of this book is as follows: Part One sets the stage; in Part Two – *Uncertainty, decision-making and innovation* – we discuss what is meant by planning, examine some theories of decision-making and look in detail at the complexities and uncertainties of the innovative decision process and the influence on this of the decision environment. Part Three deals with *uncertainty and the innovator*: In chapter 4 we report on the manner in which innovative decisions were taken in our case study firms and the role of the innovating groups in these decisions. In chapter 5 we suggest that the rational choice model of the decision theorists excludes powerful political elements which exert a major influence on technological decisions where the interests of a number of different groups have to be reconciled. We give a case study illustrative of these political influences.

In Part Four – *Uncertainty and the user* – we examine those factors which influence the way innovators define their role, the values which determine their behaviour in this role and the consequences of this role definition for the users and recipients of innovation. In chapter 6 we show how the role of the expert is subject to constant change and modification as he is affected by change and uncertainty in his own environment, particularly technical change. Chapter 7 looks at the problems associated with getting an agreed definition on the role of the innovator between computer specialists and user departments and shows some of the difficulties arising from an inability of computer specialists and users to agree on what each should be doing and to appreciate each others knowledge strengths and limitations. Chapter 8 considers political aspects of the specialist user relationship.

Many present and past members of the Manchester Business School Computer Research Unit have been actively concerned with every aspect of the project. Three in particular were responsible for collecting, analysing and writing up the research data set out in the book. These were Susan Gates, Malcolm Rigby and Krystyna Weinstein, who provided a number of the case studies from which our conclusions are drawn. The two authors also did a considerable amount of the fieldwork associated with the project. Enid Mumford would like to stress that in so far as this book contains any useful developments in organisational theory – which the authors hope is the case – these come mainly from Andrew Pettigrew, who has already published many of his ideas in *The Politics of Organizational Decision-Making*, Tavistock Publications, 1973. The manuscript was typed by Emily Stephenson, Secretary of the Computer Research Unit. The authors are most grateful for her efficient assistance.

Acknowledgements

We are indebted to the following for permission to reproduce copyright material:
American Accounting Association for an extract from an article 'Sensitivity Analysis in Decision Making' by Rappoport in *Accounting Review* July, 1967; Elsevier Scientific Publishing Company for an extract from *Policy Sciences Dilemmas in a General History of Planning* Vol. 4, 1973; The Institute of Electrical & Electronic Engineers for an extract from a working paper, Sloan School, entitled 'A Model for the Description and Evaluation of Technical Problem Solving' 1968 by Frischmuth & Allen; Institute of Management Sciences for an extract from 'A Heuristic Model For Scheduling Large Projects With Limited Resources' by Jerome D. Weist in *Management Sciences* Vol. 13, No. 6. © 1967 All Rights Reserved; McKinsey & Company Inc. for an extract from an article 'Unlocking the Computer's Profit Potential' in *McKinsey Quarterly* Vol. 5 No. 2, 1968; Organisation for Economic Co-operation & Development for extracts from 'Perspectives of Planning' by Ozbekhan edited by Jantsch 1969; The University of Chicago Press for an extract from an article 'The Cybernetic Analysis of Change' by Cadwallader from *American Journal of Sociology* Vol. 65, 1969.

We regret that we have been unable to trace the copyright holder of an extract from an article 'A Descriptive Model of the Intra-firm Innovation Process' by Knight in *Journal of Business* Vol. 40 No. 4 1967.

PART ONE

Background and definitions

1

Innovation and the generation of uncertainty

This book is about an aspect of innovation that has not yet received a great deal of attention in social science and management literature, the relationship between innovation and uncertainty. We shall argue that in order to respond to external pressures, that is to changes in the product or labour markets or in the technological environment, the firm introduces some form of innovation. It does so because it believes such a response will make it better able to cope with new demands by making it more adaptive, or more efficient, or more in tune with what its clients want. But in responding to external uncertainty by introducing new methods or products and services the firm will generate a great deal of internal uncertainty for the people who have to initiate and manage the changes.

We look at a form of innovation which can have dramatic effects on the structure and administration of a firm and will require major shifts in attitudes and behaviour patterns on the part of employees at every level: namely, the introduction of large-scale computer systems. We show how the implementation of such systems produces many different kinds of uncertainty for the firm – all of which have to be coped with if the change is to be successfully introduced. The very fact of devising means for coping with uncertainty generates additional uncertainty, often of a psychological kind, and this in turn has to be tackled if undue stress is to be avoided in the change process.

The firm thus finds itself in a looping or spiralling process which is not easy to handle. As it deals with each level of uncertainty so it produces a new level of a different kind. Innovation, by its very nature, creates uncertainty for the decision takers, the planners, the implementers and the users of the new system – who may or may not be all the same people. These groups have to cope psychologically with this uncertainty if they are to tolerate it, and this in itself requires some innovative behaviour, although of an administrative or political rather than a technical kind. This coping behaviour has consequences which, in turn, generate further uncertainty. And so the process continues until either the innovating group has become expert in handling

3

uncertainty and responds quickly and easily to it, or the innovation is implemented, becomes operational, and the situation is stabilised.

We examine the concept of uncertainty as set out in the management and social science literature and relate this theory, together with some theory of our own, to a number of case study situations: three firms and a government department, which all introduced large-scale computer based work systems into new areas of their activities.* We were able to spend periods of two to five years in each of these organisations observing the kinds of uncertainty they experienced with this form of innovation and noting their methods for coping with it.

Let us start by defining our terms. We are devoting our attention to *innovation* which is introduced in response to some recognised stimulus in the external or internal environment of the firm. We are looking, in particular, at one consequence of innovation, its capacity for generating *uncertainty* which the firm and its members then have to cope with. We must therefore make explicit what we mean by the terms, *innovation, environment* and *uncertainty* when we use them in this book.

The concept of innovation

By innovation we mean doing something new; that a person, group or organisation has not done before. K. E. Knight (1967) points out that the word 'innovation' has value overtones. It implies doing something 'good' rather than 'bad'. We accept his definition that '*an innovation is the adoption of a change that is new to the organisation and to the relevant environment*'. Innovation is a well researched subject and Knight has described how different branches of the social sciences have been interested in particular aspects of it. Psychologists have paid much attention to the beginning and end of innovation – to creativity and to change in an individual's behaviour and beliefs in response to innovation. Economists have focused on the macro implications of introducing new developments, and have considered the effect of innovation upon such things as competition and economic growth. Sociologists have concentrated upon technological innovation and its effects on the work, attitudes and needs of employees. They have also studied the management of change: the ways in which change can be facilitated and resistance to change overcome.

Until recently none of these groups paid a great deal of attention to the processes and dynamics of change: how an organisation actually shifts from state A to state B, and the problems, particularly those in the sphere of interpersonal relationships, which it encounters and solves in doing so. Large-scale change can be very complicated in that many different kinds of problems have to be handled and many different competences acquired before an innovation is assimilated and becomes operational. Coping with the uncer-

---

* In order to avoid having to refer to three firms and a government department, we shall in future refer to four *firms*.

tainty of this complex, dynamic and long-drawn-out situation may place a considerable strain on interpersonal relationships with the result that emotional and social factors are likely to play a very important part in the behaviour of both the innovator and the user of the innovation. These may affect every aspect of the 'changing' process from the first feasibility study to the planning of the change and its implementation. Some individuals and groups will see the unfreezing of the normal situation caused by the change as an opportunity for furthering their own interests through the acquisition of more responsibility, power and prestige; others will feel threatened and try to protect their present position. The first reaction will generate uncertainty for those who feel that they may lose out as a result of this acquisitive behaviour by another group. The second reaction will generate uncertainty for the person who is trying to protect his own interests for he may become fearful that he cannot succeed in doing so. We believe that any major technical change such as the introduction of a large computer system will be as greatly influenced by psychological and political motivations as it is by technical factors. In this kind of change situation the technical aspects of innovation are usually far easier to deal with than the human aspects.

Cadwallader (1969) has produced a number of cybernetic propositions determining the presence, absence and nature of innovative processes in large organisations:

1 The rate of innovation is a function of the rules organising the problem-solving trials (output) of the system.
2 The capacity for innovation cannot exceed the capacity for variety or available variety of information.
3 The rate of innovation is a function of the variety and quantity of information.
4 A facility mechanism for forgetting or disrupting organisation patterns of a high probability must be present.
5 The rate of change for the system will increase with the rate of change for the environment.

A similar set of sociological propositions might be,

1 The rate of innovation is a function of the ability of the people concerned with it to accept and to adjust to it.
2 The capacity for innovation cannot exceed the knowledge and learning ability of the people concerned.
3 The rate of innovation is a function of the variety and quantity of information and of the willingness of people to search out and use this information.
4 People must be willing and able to change old ways of doing things.
5 The rate of change for the system will increase with the rate of change for the environment providing the people in the system are prepared to recognise and respond to the environment.

Cadwallader's propositions relate to the properties of cybernetic systems;

but these systems like all others in the world of man, contain people and are therefore subject to the influences of psychological needs and of the ensuing political behaviour. For example, how you solve a problem (the rules) and how soon you solve a problem (the rate) are bound up with people's willingness to solve the problem – whether they perceive it as in their own interests to do so. The cybernetic proposition has little meaning unless the sociological proposition is allied with it. Similarly, the validity of the relationship between capacity for innovation and availability of information has the intervening variable of the competence of people to handle this information. The relationship between rate of innovation and the amount of information available is influenced by the willingness of people to seek out and use this information. They may prefer not to do this for political and psychological reasons. The facilitating mechanism for disrupting organisational patterns may be *motivation* to do so rather than any non-human factor. There will only be a relationship between rate of change in the system and rate of change in the environment if the system is responsive to its environment: many systems are not, because the people in the system do not want to change their ways of doing things.

We are trying to show here that every aspect of innovation has its psychological, sociological and political elements and these must be recognised and understood if the processes of innovation are to be understood. We therefore concentrate on these elements in this book.

The nature of environment

'Environment' is a general term which is used in many different ways. We talk of a social environment, a technical environment, a product market environment etc. referring to the characteristics of that which surrounds and influences a particular system and which the system, in turn, acts on. Ozbekhan (1969) tells us that the environment of man is his entire experiential milieu. This includes nature in all its dimensions, society, the institutions within society and so on. It also encompasses the intangible aspects of experience which we call culture, way of life, and all manner of informal relationships. Environments have certain characteristics; they have boundaries and incorporate the notion of feedback. Thus if we take a firm as one example of a system, its technical environment would be all available technical knowledge relevant to its operations. Some of this knowledge will impinge on the firm, causing it to change its behaviour. This change may then have an impact on the environment causing the nature of the technical environment in which the firm is operating to alter in some way.

Most human social systems are open systems – that is they interact with their environments, receiving inputs from them and in return sending forth outputs into the environment. By input we mean an event occurring in the environment that alters the system; by output, any change caused in the environment by the system.

One of the most important environments in which a firm operates is its product market. If it cannot find customers and sell them the goods it makes it will go out of business. All successful firms are highly responsive to the demands and pressures of their product market environments. However the firm also has an internal environment. For the people it employs, its structure, procedures and value system is their environment. They will both respond to and influence this. The external and internal environments of a firm are constantly interacting with each other and in doing so they will affect the social system – the network of human relationships which enables production activities to be carried out and which also provides individuals and groups with psychological support and satisfaction.

A study of one kind of environment of particular relevance to our discussions in this book has been made by Duncan (1973). He has looked at the decision-making environment and analysed how different dimensions of this affect the amount of difficulty and uncertainty experienced when decisions are taken. He defines the decision-making environment as 'the totality of physical and social factors that are taken directly into consideration in the decision-making behaviour of individuals in the organisation'. In later chapters we pay a great deal of attention to the nature of the decisions taken when introducing large-scale computer systems, to the problems encountered in taking innovative decisions of this kind, and to the changes that occur because such decisions have been taken. Duncan's work is therefore of considerable value in directing our attention to particular problem areas.

Duncan considered two dimensions of the decision environment. The simple-complex dimension which incorporates the number of factors taken into consideration in decision-making, and the static-dynamic dimension – the degree to which these factors remain the same over time or are in a process of continual change. He finds that dealing with a large number of decision variables is less troublesome than taking decisions in an environment that is constantly changing. This is an interesting and important point for a feature of the introduction of large-scale computer systems is the long time cycle, with periods of years between the decision to innovate and final implementation. During this time both internal and external decision environments are likely to alter. Changes in the product market environment may cause the firm to rethink its computer plans, or changes in the technological environment may bring new technical opportunities of which the firm wishes to take advantage. The internal decision environment Duncan (1973) defines as 'those relevant physical and social factors within the boundaries of the organization or specific decision unit that are taken directly into consideration in the decision-making behaviour of individuals in that system'. We shall show that many of these factors are political in nature and that the fact of taking complex decisions in an uncertain environment provides the opportunity for the expression of political behaviour, while the fact that political behaviour occurs generates additional uncertainty for the participants in the decision-making process. An important skill in anyone involved in decision-making is therefore the ability to tolerate and manage uncertainty in both the internal and external environments.

Successful innovation requires that an organisation is able to spot when a change has occurred in its external environment, to which it needs to respond if it is to maintain or increase its ability to survive. The organisation must therefore constantly gather and process information on what is happening outside. Duncan (1973) argues that uncertainty is generated for decision makers because they are unable to obtain this information in a precise and accurate form. There is a lack of information on what factors in the external environment are most relevant to the decision that is being taken. For example, a decision may be strongly influenced by perceived changes in the firm's product market; the fact that even more dramatic changes are taking place in the labour market may pass unnoticed. Yet the labour market may be more critical to the firm's survival at that time than the product market. The decision makers will also experience problems because they cannot know how much the firm will lose if the decision they make is a poor one. This places them in a very high risk situation and makes them specially vulnerable if the decision they are making relates to an innovation that has not been attempted before. Again there will be problems because it may be impossible to weight environmental factors in terms of their ability to affect the success or failure of the decision that is made.

Similar issues could be set out for the internal environment. The industrial relations climate and structure is always a factor likely to introduce uncertainty into any decision to embark on a major innovation. It may be neither customary nor easy to find out how the employees and their unions will react. The history of technology is full of stories of advanced machines rusting away unused because of industrial relations problems. Internal groups will always attempt to influence decisions which they perceive as affecting their interests. They will take a particular position in relation to a proposed change because they look at it primarily from their own point of view, and because their interpretation of the consequences of the change is biased by the job they do within the organisation and the value system of the occupational group to which they belong. The history of relationships between parties in the decision process may also be crucial. In many capital investment decisions the past weighs a heavy hand. Figure 1.1 shows the influence of the external and internal environments on the decision process.

Randall (1973) points out that all organisations face some hostile forces in their environment. The organisation, in order to survive, must be aware of these and take steps to reduce or neutralise them. As a rule it does this by securing support for its goals and objectives from the environment. It may, for example, seek to acquire a large group of loyal and well satisfied customers. But, in order not to endanger this support, it may become conservative and unable to spot when an important change is occurring in the environment. The old and loyal customers may be dying out and either not being replaced at all, or being replaced by a younger group with different needs. An organisation

Fig. 1.1 Factors in the internal and external environments which influence the choice of goals and solutions

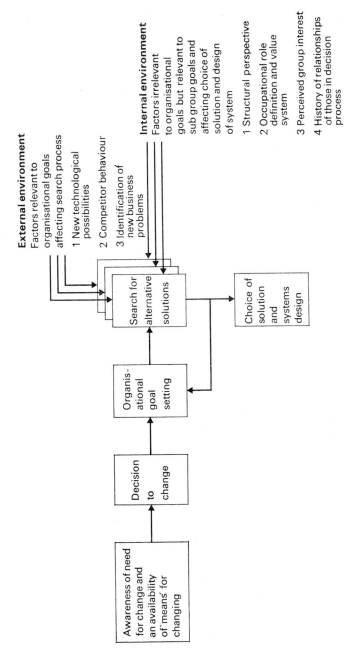

is in a favourable position if it can find what Randall calls a 'policy space'. This is a space in whatever market it serves where it is not subject to competition from other groups. It is also in a favourable position if it can get 'environmental support' – the development of a clientele wishing to purchase its product or service. An important reason for introducing computer systems is often to increase environmental support. The firm wishes to prevent its loyal customers from deserting and to gain new ones by improving its service to customers. It sees a computer system as an aid to doing this.

These concepts of 'policy space' and 'environmental support' can also be applied to the internal environment of the firm and to the role and relationships of the computer department. It can be argued that the early rapid growth and power of these departments was because they were able to introduce a form of innovation that was new to the firm and helped it to cope more successfully with its external market environment. They therefore occupied a new technological 'policy space' within the firm. But although computer departments can design and introduce computer systems, they do not as a rule operate them as well. They therefore require a clientele anxious for their product and conscious of its value. It is in this area of the internal environment that the most serious difficulties have been encountered. We look at the problems of innovator–user relationships in chapters 7 and 8.

The concept of uncertainty

Galbraith (1973) has traced the history of organisational theory in recent years and shown how each new approach to understanding the way organisations function takes us on to another important area of interest. Burns and Stalker (1961) in their study of innovation in the electronics industry produced the interesting theory that the way a firm organises itself is related to the kind of market environment in which it is operating. If the market is stable and unchanging then the firm is likely to adopt a stable, well structured organisational form which they call 'mechanistic'. If in contrast, the market environment is changing or the technology which the firm uses is altering then the firm requires a greater capacity to adapt to these changes in the environment and it needs to assume a more fluid, 'organic' form.

In the late 1950s and early 1960s Joan Woodward was carrying out and publishing her research on organisation theory (see Woodward, 1965). She found a relationship between the type of production process used by the firm and its organisational structure. She argued that firms were unwise to organise themselves on the basis of classical management principles alone: they needed to ensure that their organisational form fitted the nature of their production processes. Firms which used batch, mass or process production systems should have a different organisational logic.

Harvey (1968) developed Woodward's ideas further and looked at the relationship between organisational form and the kinds of products which the firm made. He claimed that the production process was dependent upon the nature of customer demand. Thus a firm whose customers wanted unique products made to their specifications would have to have a one-off or small batch production process.

Hall (1962) looked not only at the firm's products but at the kinds of tasks that were required to make these products. He hypothesised that the predictability of the task would have a powerful effect upon organisational structure. Departments with well defined tasks would have well defined functions and a clear hierarchy of authority. In contrast departments whose tasks were less predictable, such as R and D, would exhibit these characteristics to a lesser degree, as they would need to be more responsive to the demands of task changes.

Lawrence and Lorsch (1967) identified two further dimensions of the organisational design problem; *differentiation* and *integration*. Following Hall they suggested that different kinds of organisational structure should be used for predictable and non-predictable task systems. The viable firm will not organise each of its departments and functions in the same way but will vary its organisation to suit the nature of the task. This is differentiation at a sub-task level. In order to achieve the completion of the firm's whole task, its final products and services, these subtasks must be successfully *integrated* so that each one makes the necessary contribution to the whole.

In the development of these theoretical ideas we can see an increasing recognition of the concept of uncertainty: the fact that when a firm is operating in a changing environment or carrying out tasks which have areas of unpredictability, then it has to be able to cope with uncertainty. This requires a capacity for response to feedback, that is to signals from the environment that a situation has changed as a result of an earlier action; this in turn requires an ability for fast adaptation to meet the needs of the new situation.

Lawrence and Lorsch (1967) suggest that the greater the unpredictability and uncertainty of the task, the greater the amount of information that has to be processed during the execution of the task. If the task is well understood prior to performing it, then much of the activity can be preplanned. If it is not understood, then in carrying it out more knowledge is acquired which leads to changes in resource allocations, schedules and priorities. These changes may generate additional uncertainty for the performers of the task.

Most researchers into the concept of uncertainty, including Galbraith, are particularly interested in the relationship between uncertainty and information, the latter being seen as an important way of reducing uncertainty. Duncan (1972) looks at the different approaches to uncertainty of the information theorists and the decision theorists. He criticises the information theorists for taking too narrow an approach and defining the concept in terms such as those of Garner (1962), that 'the uncertainty of an event is the logarithm of the number of possible outcomes the event can have'. The decision theorists, for example F. Knight (1921), Luce and Raiffa (1957), and Raiffa (1968) define uncertainty as those situations where the probability of the outcome of events is unknown. This is in contrast to risk situations where each outcome has a known probability. Lawrence and Lorsch take an even wider definition and state that uncertainty consists of three components,

1 the lack of clarity of information;
2 the long time span of definitive feedback;
3 the general uncertainty of causal relationships.

Most theorists have paid more attention to uncertainty arising in the external environment of the firm than to uncertainty which is a product of factors within the internal environment. Thompson (1967) is an example of this school. He sees uncertainty as the fundamental problem for complex organisations and coping with uncertainty as the essence of the administrative process. Uncertainty in the external environment arises from two principal sources. The first is generalised and is due to a lack of understanding of the relationship between causes and effects in the culture as a whole. The second is a contingency form of uncertainty in which the outcomes of organisational action are in part determined by factors operating in the external environment. The firm can thus never be sure that an action directed at the achievement of a certain end will in fact produce the desired end result.

Argyris (1972) has criticised Thompson for paying so much attention to external uncertainty that he ignores uncertainty as a problem within the firm. He points out that uncertainty which arises as a result of internal factors has been a subject for study in the past. The theories of the early scientific management school were attempts to deal with internal uncertainty by providing principles of administration which would lead to a controlled and orderly internal environment. These early theorists assumed that if the internal environment was managed correctly the firm would be tuned correctly to the requirements of the external environment, although they paid little attention to the nature of the relationship between the firm and its external environment, or to the influence of the external environment on the behaviour of the firm.

Argyris (1972) also criticises studies of uncertainty for ignoring man as an active element with a great deal of influence on the way the organisation is structured. He argues that different organisational forms are not solely a result of product market influences, of technology, or of task unpredictability. They may be there because an individual at a senior level in the firm has made a value judgment and said 'I think we should organise the firm in this way', and his influence or political skills have been such that he has got other people to agree with him. Argyris maintains that managements are increasingly and with much deliberation causing turbulence and uncertainty within their organisations. They are doing this because they wish to remove some of the firm's dry rot and to increase its flexibility. Managements do not look at their product markets or their technology and say 'because we have this kind of environment we must organise ourselves in this way'. They see only large numbers of mutually dependent variables and their own task as to so manipulate these variables that the organisation is not surprised by environmental change, that it is constantly alert to the need for change and that it can manage the transition to a new system in such a way that survival is possible. Argyris therefore would like more attention paid to management values as an important influence on the way uncertainty is tackled by a firm. We show in this book how the values and attitudes of the innovators in our case study firms were amongst the strongest influences on the outcomes in the decision processes.

Duncan (1972) gives Argyris some support in this point of view. He says that if we accept the position that an organisation has no properties aside from the way people perceive it, then we need to identify more clearly how individual differences affect perceptions of organisational properties (see Hunt, 1968). What one individual sees as an internal environment full of stress and uncertainty, another individual may interpret as benign and ordered. What one individual sees as a tightly structured, mechanistic form of organisation, another may regard as adaptive and organic. Therefore, argues Duncan, research should examine the impact of individual differences on the perception of uncertainty and the complexities and dynamics of the organisation's internal and external environment. This research would help to develop a more comprehensive contingency theory of organisations, that is, a greater understanding of how organisations can develop the structures most appropriate for their needs. Duncan suggests that most contingency theories available at present are onesided. The researchers we have mentioned (Burns and Stalker, Woodward, Harvey, Hall and Lawrence and Lorsch) although providing great insights into why particular organisational forms arise, focus on the characteristics of the external environment or of the technology or task situation while ignoring the equally important contingent factor of individual differences among organisational members.

Cyert and March (1963) point out that organisations not only learn to cope with uncertainty, they also try to avoid it altogether. They do this by concentrating their attention on immediate and short-run problems which can be dealt with through decision rules, and pay little attention to long-term strategies. They may also attempt to create a negotiated environment so as to reduce the possibility of external uncertainty. They create industry traditions, price rings and uncertainty absorbing contracts with competing firms. They achieve a reasonably manageable decision situation by avoiding planning where plans depend on prediction of uncertain future events and by emphasising planning where plans can be made self-confirming through some control device. If coping with uncertainty cannot be avoided it is coped with through short-run feedback (crisis management) in which new events are not dealt with until they happen. Unfortunately none of these courses of action is appropriate for any large-scale innovation of the kind we describe in this book. Here the long time cycle of the change process means that long-term problems must be given a great deal of consideration. A negotiated external environment is quite impossible because a major source of uncertainty is the

fact that this environment changes very rapidly, particularly in its technological aspects. Crisis management may be used at certain points in the
decision and implementation processes, but this is likely to prove expensive
and to raise the overall level of uncertainty.

The modern firm must become expert in dealing with uncertainty that
arises in both its internal and external environments. It must recognise that
the management of external uncertainty will almost certainly generate internal uncertainty, particularly if the response to change in the external
environment is some form of innovation such as a major computer system.
The nature of the uncertainty must be understood, its consequences known,
and means for using it as a healthy stimulus rather than a major threat must
be developed. This is the challenge for management.

Figure 1.2 shows some of the areas of internal uncertainty that have to be
dealt with when introducing a computer system.

Ways in which organisations cope with uncertainty

In the following chapters of this book we consider the different kinds of
internal uncertainty that have to be managed when major innovations are
introduced and the means that are used to cope with this uncertainty. We
believe that uncertainty will be generated for a firm because of an absence of
required information, because of a need to plan for outcomes that have large
elements of unpredictability, and because any major change is likely to lead
to political behaviour as participants in the change process use the unfreezing
of the normal situation as an opportunity for competing for scarce resources
and furthering their own interests.

The literature on decision-making and uncertainty suggests that organisations have a number of well established methods for handling internal uncertainty. They use experts to provide the knowledge on which decisions will
be based, they use tested planning procedures as a means for anticipating
uncertainty and providing guidelines for dealing with it. They have ways of
obtaining, ordering and controlling information, again as an aid to making
decisions. We would also argue that the political behaviour that erupts, often
with considerable violence, during periods of stress and change is an attempt
to control uncertainty by individuals and groups. If people are able to increase their personal or group power during the change process they are in a
much better position to influence events. Alternatively, if they are able to
prevent other individuals or groups acquiring power then this places them in
a better position to protect their own interests. We are particularly interested
in political behaviour as a means for coping with uncertainty as this has received little attention in the literature.

The use of the expert

Decisions on when and how to use computer systems tend at present to be
dominated by technical experts who are in staff rather than line functions.

Fig. 1.2 Innovation and the generation of uncertainty

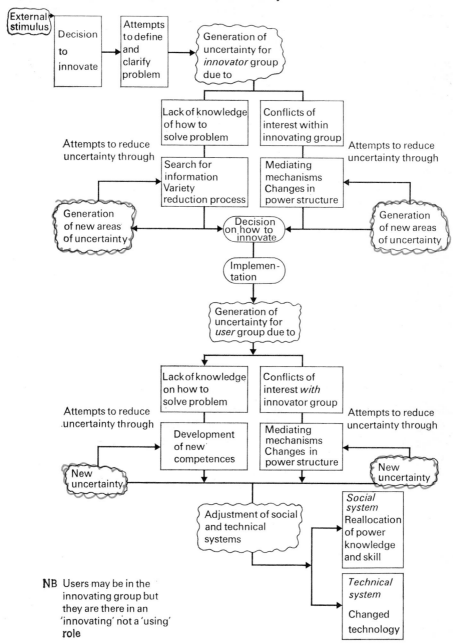

NB Users may be in the
innovating group but
they are there in an
'innovating' not a 'using'
role

These computer professionals are used by their firms as sources of knowledge and information; as people competent to carry out complex search processes to establish what technical alternatives are available and to evaluate these alternatives in terms of their ability to meet the firm's needs. They thus have a filtering and sifting role, removing uncertainty from senior management by carrying out preliminary investigations and making recommendations for action. In chapter 4 we describe in some detail how our computer experts carried out this information collecting and screening process.

Although the use of experts protects management at the top of the firm from the uncertainty of making decisions in areas where they have little knowledge; the computer expert may increase the uncertainty faced by managers and employees in those areas of the firm where new computer systems are to be introduced. They may feel that his technocratic bias will lead him to design a computer system which meets technical but not human needs. They may fear that he has little understanding of human relations and democratic principles and will impose a computer based solution on them which they do not fully understand and do not believe themselves competent to operate. Alternatively they may be under stress, not because a system is being imposed on them, but because they are expected to design their own system even though they do not have the necessary experience and knowledge to enable them to do this. In chapter 7 we provide a case study example of some of these problems of user–expert interpersonal relationships.

Planning

Planning can be an aid in meeting short-term uncertainty by providing guidelines on how new objectives can be set and reached. Once the planning timescale lengthens, however, rigid plans can be dysfunctional because they do not cater for the need to alter both means and ends in response to changes that have occurred in the organisational environment. There is then a need for adaptive planning in which plans react to new uncertainties in the environment. Ozbekhan (1969) tells us that:

> Planning is adaptive to evolution in the environment only in so far as continuous exchanges between it and the environment permit it to operate in an adaptive mode. More often, however, because of growing rigidity and institutionalisation, plans become non-adaptive and tend to lose touch with surrounding evolutionary trends.

He believes that the *raison d'être* of any planning is to change the environment in a manner that is smooth, timely, and orderly so that a dynamic social evolution that is in line with our ideas of organised progress can be achieved.

Ozbekhan is very much aware of the human aspects of the planning process. He tells us that planning is not merely a set of abstract concepts, projections of trends or decision procedures; it is also people. People who are working in conditions of considerable uncertainty, in which they may be uncertain about both their own private futures and the future of the function they are carrying out. Because the objective of planning is 'change' it is likely to be a threat. For change often means a reallocation of resources and a consequent alteration in relationships and in some of the perceived rewards of institutional life: status, power and influence. Institutions often try to resist these consequences of change by strengthening their administrative characteristics; those factors which assist the continuity of existing relationships and the stability of the organisation.

This can lead to a pathological situation in which the firm, while needing change if it is to interact successfully with its environment and hiring experts and setting up planning units to achieve this change, at the same time organises itself to resist change. This increases the uncertainty of the environment in which the planners are operating. With new computer systems conservatism of this kind can occur in the user departments, and we provide some examples later. But it can also occur in the innovating groups themselves so that these groups may be change agents for others but highly resistant to change themselves. In chapters 4 and 5 we show how many of the actions of the computer specialists were related to attempts to maintain personal and group security in an uncertain situation.

Innovative planning of the kind we describe below seeks to achieve very fundamental change. It seeks to legitimate new social objectives and values and to translate these into new institutional arrangements and concrete action programmes. Friedmann (1967) describes innovative planners as entrepreneurs who try to get the largest amount of available resources for their projects. They are more interested in mobilising resources than in allocating these in an equitable manner between other groups in the firm. They are not interested in the gradual modification of existing procedures, they seek a major change which is quick and dramatic. Their problem is to identify the critical points in the organisation where change can be easily and successfully introduced.

Rittel and Webber (1973) make the interesting point that planning reality bears little or no relation to planning theory. The theorists describe the functioning of an ideal planning system in terms like the following:

> It is seen as an ongoing, cybernetic process of governance, incorporating systematic procedures for continuously searching out goals; identifying problems; forecasting uncontrollable contextual changes; inventing alternative strategies, tactics and time sequence actions; stimulating alternative and plausible action sets and their consequences; evaluating alternatively forecasted outcomes; statistically monitoring those conditions of the publics and of systems that are judged to be germane; feeding back information to the stimulation and decision channels so that errors can be corrected – all in a simultaneously functioning governing process.

This is the modern-classical model of planning. And yet we all know that this approach to planning is unattainable and may even be undesirable. A major defect of the model is that it ignores many of the 'real' issues that affect the planning process. These, as we have indicated, stem from personal and group interests and will affect the way in which organisational interests are pursued. To quote our source: 'The expert is also the player in a political game, seeking to promote his private vision of goodness over others.'

Fig. 1.3 A model of the ideal planning process

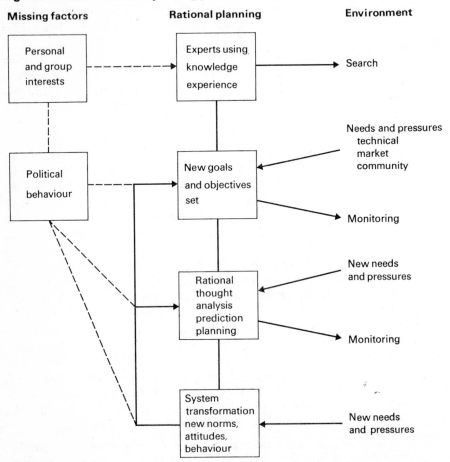

Rittel and Webber (1973) point out that the problems with which innovative planners are having to deal are becoming increasingly difficult. These contain elements related to powerful human needs which have been ignored in the past. They are therefore social problems which can never be solved but only re-solved over and over again. The ability to understand a really difficult problem may depend on a knowledge of the different ways in which the problem can be solved. Unless one knows how to solve it one cannot know what information is required to understand it, and so a vicious circle is introduced in which many problems become insoluble. The planner terminates work on these problems, not because he has arrived at a logical solution, but for reasons which are external to the problem; he runs out of time, or money, or patience. Computer problems are of this kind. A perfect solution can never be attained because the nature of the available alternative solutions is constantly changing. This is due to the rapid development of the technology. As computers improve so better means for solving business problems appear on the scene.

The acquisition and ordering of information

One way of controlling uncertainty is through the gathering and processing of information, and this is an essential function of any group responsible for innovation. Cyert and March (1963) tell us that when a group searching for information comes to a dead end and the members cannot find the information they require, then this fact has both short- and long-term consequences. In the short term it means that relevant information is not available where it can be used and decisions are taken on the basis of incomplete information. Difficulties arise here from the fact that it is not always possible to know what is relevant, and we show later how technical experts focus on technical information and do not search for other kinds of information that may have relevance to the problem they are trying to solve. Even if certain kinds of information are known to be relevant they may not be accessible, but buried in other parts of the organisation or its environment. These dead ends lead to a decision strategy that involves extensive use of what are essentially contingent decisions. Decisions are used as devices for learning about their hidden consequences. A decision is taken and is quickly altered if an outcry suggests that it is the wrong one.

All planning and decision-making associated with innovation requires that information shall be processed and Galbraith (1973) suggests that the greater the uncertainty of the task undertaken, the greater the amount of information that must be processed in order to arrive at a successful conclusion. Uncertainty is not the only factor that generates a high need for information, however, interdependence is another factor. If the behaviour of one department affects that of another, then communication is essential. Galbraith describes the different ways in which organisations handle these two factors of uncertainty and interdependence through the way they organise their information systems. He sets out eight different ways in which a firm can assist the processing of information: vertical information systems, lateral relationships, liaison roles, task forces, teams, integrating personnel, integrating departments and matrix departments. Some firms attempt to increase their ability to handle information through operating on the existing formal, vertical information system. Attempts are made to reduce the number of exceptions flowing up the information hierarchy and in order to do this planning cycles are shortened. Long-term planning is avoided because this involves dealing with a great deal of uncertainty and this type of plan is subject to a rapid decay rate. Additional information-processing and decision-making that cannot be avoided is handled by adding assistants and clerical personnel to different levels of the hierarchy.

A second way of increasing information processing capacity is through the establishment of lateral relationships and the making of joint decisions. Managers in strongly hierarchical organisations often find this form of relationship difficult. It forces them to behave cooperatively with peers without the security of an authority relationship. Liaison roles are another means for easing communication. These are specially created roles to handle communication between departments. They are not uncommon when computer systems are introduced, a member of the user department being given responsibility for liaising with the computer specialists. The problem here is retaining neutrality. It has been found that members of user departments given this role become more identified with the interests of the computer department than with those of their own department. When a problem is very large and involves, say, seven or eight departments, liaison through the direct contact of interested parties may become impossible. It may then be necessary to create a task force made up of representatives of all these departments. This is usually a temporary group with some full-time members and some part-time and it uses a group problem-solving approach to decision-making. But, as the task becomes less predictable and more uncertain, so temporary teams become less efficient. It may then be preferable to form permanent teams with responsibility for certain kinds of problem-solving. Management services and research and development departments are examples of this kind of permanent problem-solving group.

Galbraith suggests that as uncertainty grows and more information has to be handled, so decision-making moves downwards and the amount of discretion located at lower levels of the organisation increases. The organisation now often tries to raise the level of decision-making upwards again and creates a number of new roles at a senior level with responsibility for integration. Roles associated with corporate strategy, long-term planning and development are examples. If these roles work well then the people holding them are likely to be allotted more power; they are given subordinates and may find themselves heading a new department with an integrating function.

The last approach for dealing with uncertainty is the matrix organisation. In firms which have to deal with a continual high level of uncertainty a fluid organisational structure emerges, with a high capacity for joint problem-solving and shared responsibility. Matrix organisations have dual authority structures. Traditional functions such as marketing, production and R and D are to be found there, but as new problems arise so problem-solving groups are formed representing all of these departments. These groups stay together until the problem is eventually solved.

Galbraith's argument is that coping with uncertainty requires effective information collection and processing. The more uncertain the task, the more information is required and the more pressing the need to be able to handle this competently. In order to do this the firm must design organisational structures which are effective information processors. If it cannot create these it must reduce the amount of information which it has to handle. It can do this by reducing the length of the planning cycle or by introducing some slack resources to lessen the tightness of the interdependence between departments. Alternatively, it can decentralise and thus bring decision-making down to lower levels in the organisation, thus spreading the information-processing load. In chapter 4 we examine how our firms dealt with the problem of processing information as a preliminary step to the making of decisions.

Political behaviour as a means for controlling personal or group uncertainty

Innovation is inextricably bound up with politics and power. Innovating groups whose activities are sanctioned by senior management tend to be given power in the sense that they are allowed to acquire the resources necessary for innovation. These resources include intangibles such as permission to influence and command other people as well as material resources such as new equipment. Giving one group power may remove power from other groups, or may lead to confrontation with these groups if they fight to retain their power. These conflicts are some of the most stressful aspects of innovator–user relationships when new computer systems are introduced and add greatly to the uncertainty of the change situation. Firms may find themselves in a state of positive feedback at this time. The need to cope with the uncertainty of change leads to political behaviour; the fact that this political behaviour occurs produces more uncertainty, and so on. With innovation the power of the expert is often a significant factor, for he is the man who understands the new technology and can get it working and put it right if it breaks down. Crozier (1964) describes how technical experts try to associate an element of mystique with their knowledge. They fight off attempts to rationalise it and to make it readily communicable to others, as this would reduce their power and influence. He says:

> The power of A over B depends on A's ability to predict B's behaviour and on the uncertainty of B about A's behaviour. As long as the requirements of action create situations of uncertainty, the individuals who face them have power over those who are affected by the results of their choice.

Crozier presents two hypotheses concerning the relationship between uncertainty and power:

1 The more complex and dynamic the system of power relationships and of bargaining, the more likely are social controls to be directly and consciously enforced by management.

In other words he suggests that management will attempt to control social relationships in an uncertain situation to ensure that conflict does not get out of hand.

2 An organisation which is used to change will accept change readily since pressures to eliminate uncertainty will not be held in check by the resistance of well entrenched groups.

This kind of organisation will understand uncertainty and know how to deal with it. It will not tighten its administrative procedures in an attempt to preserve the *status quo*.

Innovation usually implies some reallocation of scarce resources. It provides an opportunity for groups and departments to gain control of assets they did not possess before. This in turn implies some control over social pressures and makes it an intensely political activity, as we show in chapter 5; political activity requires political strategies and we analyse those that occurred within our innovating groups and between innovator and user groups.

Conclusion

In this chapter we have tried to define the terms 'innovation', 'environment' and 'uncertainty' as these are the subjects discussed in this book. We have also briefly indicated some of the ways in which firms attempt to cope with the uncertainty of major innovation. They set up groups of experts, they introduce planning procedures and develop methods for handling the mass of information required to take decisions in a new area. However, the uncertainty which is a product of a fluid change situation and the risks inherent in major innovative decisions generate political behaviour directed at protecting individual and group interests and at increasing power and influence through the mobilisation of newly available resources. This political aspect of innovation pervades all the decision and change processes and yet receives little attention in the existing management and social science literature. We therefore pay considerable attention to it here.

PART TWO

Uncertainty, decision-making and innovation

2

Planning prescription - its contribution to coping with uncertainty

Planning has been defined by Newman (1951) as 'deciding in advance what is to be done'. This is a nice neat definition but so broad that it covers almost every human activity. Shafer (Shafer *et al.*, 1967) narrows the definition somewhat by suggesting that planning is 'the process of formulating goals and developing commitments to attaining them; a process undertaken to some extent by all organisations and individuals'. Both of these definitions focus on the end product of planning – the achievement of some kind of goal – but place little weight on the processes for reaching that goal: the means by which an organisation gets from an existing state A to a desired state B and the nature of the environment that has to be travelled through in order to reach B.

Planning is perceived by many writers to be a result of an organisation's recognition that it must set new goals if it is to cope with *external* environmental uncertainty such as new technological developments, or changing product or labour markets. But there is little explicit reference in the planning literature to the nature of the *internal* organisational uncertainty that must be understood, ordered and controlled if goals are to be achieved, and if these goals are to solve not only the problems which were defined when the planning process was started at state A, but also the new problems and needs which will almost certainly have arisen during the timelag between starting at A and arriving at B.

Planning is a way of controlling uncertainty. It usually begins as a response to some recognised external pressure for change which influences the organisation to set new objectives. But if it is to lead to what the organisation regards as a successful conclusion it must cope with a great deal of internal uncertainty, much of which is generated by the nature of the planning process itself. In this chapter we examine planning prescription and attempt to establish the extent to which it recognises and defines the nature of this uncertainty.

A great deal of the literature dealing with planning practice concentrates on setting down guidelines and principles for action. Two assumptions appear to be made: first that the principles are universally applicable, irrespective of the planning context; second, that if adhered to they will work with comparative smoothness, and that there will be no informal influences associated with the planning process and perhaps generated by it, to make a smooth movement from A to B difficult.

Emery is one writer who presents a very neat, structured view of the planning process. He sees planning as essentially hierarchical in character with broad planning policy generated at the top, more detailed policy at middle management level, with the details worked out and implemented at the operational level. Emery (1969) believes that *formal* planning 'involves the explicit evaluation of alternative courses of action, selection of one of the alternatives for execution, and formal communication of the decision to interested parties throughout the organisation'. He suggests that a great many activities fall within this definition, including the setting of organisational goals, the design of organisational structures, the selection of resources, the specification of policies and procedures, budgeting, systems design, product design and detailed scheduling. (See also Goetz, 1949.)

This view of planning sees it essentially as a management activity and Sackman (1971) criticises Emery for paying virtually no attention to the contribution of human elements, such as problem formulation, consensus, creativity, and problem solving, or negotiation and compromise.

Shafer appears to have greater recognition of the uncertainty and complexity of the planning process when he makes the point that 'planning' must be distinguished from plans (Shafer *et al.*, 1967). Planning is a process of change and innovation, a process in which the planner himself is not 'programmed' but is in a flexible and creative role. Planning is *intended rationality*, but in practice it may contain strong irrational elements. Its objective is imaginative innovation, and goals should be associated with creative and not routinised changes. But although Shafer takes a less mechanistic view of planning than Emery he still focuses on planning goals and pays little attention to the importance of the planning environment in the achievement of these goals.

Anthony (1965) believes that in order for planning to be understood it is necessary to distinguish between two aspects of planning which he sees as fundamentally different. These are *strategic* planning and what he calls *management* planning. He defines strategic planning as policy formulation and goal setting for the organisation as a whole; a process which involves deciding on company objectives, choosing the resources necessary to achieve these objectives and the policies which are to be used to govern the acquisition, use and disposition of these resources. (See also Mumford, 1968.) Strategic planning is therefore very much a staff and top management process. Anthony defines management planning and control as a type of planning concerned with the administration of the enterprise – for example, bringing new systems of work into operation and formulating personnel practices to meet departmental needs. He sees these localised planning functions as line activities.

Here again we have a prescriptive approach to the planning process with analysis confined to a separation of planning activities in such a way that they can be allocated to different groups within a firm.

If we consider the introduction of computer based systems in terms of Anthony's approach then strategic planning will cover the determination of electronic data processing (EDP) objectives for the company as a whole while management planning is the design and implementation of the new computer based work systems.

Many of these writers are aware that uncertainty in the planning environment is something that must be catered for, although they do not make the nature of this uncertainty explicit. There is a recognition of a need for adaptive planning, and stress is laid on the fact that few objectives can remain unchanged for long periods. Most early goals will have to be reformulated as conditions alter both within the company and in its external environment. There are recommendations that firms should set up established procedures which will enable top management to identify quickly new or changed conditions and to restructure objectives in the light of these. Successful planning is seen as flexible and adaptive, containing mechanisms for reformulating goals as circumstances alter and earlier objectives become inappropriate for new needs. Some writers recommend control systems that incorporate effective monitoring and feedback mechanisms. Emery (1969) claims that successful planning

> calls for a control system that senses actual events and compares them with the predictions on which current plans are based. If a significant deviation occurs, the control system should signal the need for new planning that takes into account the most recent available feedback information about the environment.

Uncertainty is therefore recognised and catered for, in theory if not in practice, for details of the nature of these sensing mechanisms are not provided, but it is not defined and described.

Once top management has determined its EDP objectives, planning theorists argue that it must now engage in a strictly rational planning and decision-making process. Emery suggests that this should involve the setting out of alternative courses of action likely to assist in achieving these objectives, and attempting to forecast the consequences of each one of these in terms of financial or efficiency payoff, short- and long-term plans, ease of introduction, impact on the structure and organisation of the company and effect on staff. At this stage management must also assess the availability and cost of the resources required by particular alternatives as a basis for making comparisons – for example, knowledge, expertise and software. Finally, once the planning group has considered the consequences of alternative courses of action it must decide on which course of action is most appropriate, given objectives, resources and constraints.

If this algorithmic approach is accepted then the planning process should be as rational and sequential as in Fig. 2.1. If the external or internal situation changes then this uncertainty can be dealt with through a rapid reformulation of goals and another journey through the steps set out in the diagram.

Fig. 2.1 A prescription for 'rational' planning – the strategic planning process

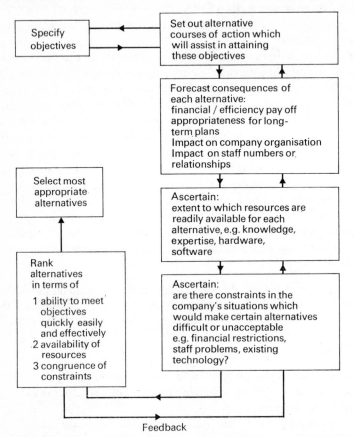

Feedback

The assumption behind this approach is that strategic planning can be a tidy sequential process carried out logically, and leading to the possibility of clear final choices. In practice, this may be worth aiming for but it is important to recognise the nature of the difficulties that will be encountered when using such an approach. Hawgood and Mumford (1970), writing about planning for the introduction of new computer systems, argue that the broad planning principles which should be aimed at when introducing these systems should be the same as those to which managers pay at least lip service when planning any organisational or technological change:

1 that decisions should be made in as rational and informed a manner as possible;

2 that any changes should be carefully prepared on both the human and the technical side;

3 that progress in design and implementation, and performance in operation, should be monitored against prior estimates;

4 that control action should be taken when needed.

But they recognise the elements of uncertainty in the automation field which appear to make the application of a coherent and rational planning strategy very difficult and very rare. For example, much uncertainty in the human relations area is generated through the awareness by those involved in the change that the introduction of the proposed computer system will have consequences both for the structure of the enterprise as a whole and for the task structure and responsibilities of affected departments. When task structures change so does the work that employees are required to do, and so may their job satisfaction.

One way of meeting uncertainty is to ignore those areas where it is most likely to arise, namely human needs, expectations and relationships. Observation of EDP planning practice shows that top management and computer specialists often seek the security of the familiar and the quantifiable and therefore focus on financial and efficiency goals while giving scant attention to the human aspects of the change – which may be both difficult to predict and to handle. This approach can lead to serious difficulties, for plans which appear to be economically and technically sound often fail for political or social reasons which could have been anticipated and avoided. K. F. Walker (1971) has provided some reasons for the narrow selection of goals associated with strategic planning, particularly when computer specialists play a major part in this process. He suggests that the latter group sees the company as a technical system designed to achieve technical and economic objectives and that they regard human beings as necessary but unpredictable elements in this system. Computer specialists attempt to achieve personal security and order by defining problems in terms of factors which they believe they can understand and control. Sociologists have argued that many managers and technologists have an engineering concept of the firm and see it in terms of information and product input and output. Human beings are weak links in this potentially efficient system and it would be preferable if they could be replaced by machine links. They do not see their firm as a fluid system in which human needs play a major role in determining attitudes and behaviour.

Planning, with or without the inclusion of human goals, is rarely as rational as planning theorists would like to suggest, and in this book we analyse some of the factors which make rationality difficult. Shafer et al. (1967) point out that rational behaviour is usually intermingled with behaviour which is spontaneous, inadvertent or random, or behaviour which is directed by tradition, previous decision or external pressure. Planning as foresight and prediction, coupled with goals, policies and action, is linked to and influenced by the specific culture in which the planning process is taking place and the set of values that predominates in that culture. The concept of planning must imply some form of commitment in the sense of an acceptance of planning objectives and a willingness to implement and operate new systems which emerge from the planning process. This implies much more than merely formulating proposals, searching out alternatives and making choices. It also implies a degree of organisational consensus on the correctness of the actions that are being taken or, at least, some means for reconciling differences in subgoals as planning progresses.

Planning is also influenced and made uncertain by the factor of time. Any major change such as the introduction of a large EDP system will have a very long time gap between the first formulation of goals and the eventual implementation of the system; in some of the computer systems referred to in this book this was as long as five years. During this period new managers, with new interests arrived in the firm and old managers left. There were changes in the power relationships of different departments and changes in the business goals of the firm as a whole. All these factors affected, and had to be recognised and catered for, by the planning process. They introduced new constraints and provided new opportunities; they also made it impossible to achieve anything resembling the logical progressive decision-making of the rational planning model.

Sackman (1971) presents one of the best methods for dealing with planning uncertainty. He recognises the problems associated with the fluidity of the planning environment and our ignorance of how to deal with this. In place of a prescriptive approach, which he regards as untenable and unrealistic, he substitutes an experimental approach:

> The foundation stone is disarmingly simple, but crucial: plans may be conceived as hypotheses, subject to empirical test and evaluation in a scientific manner. Given certain conditions, hypotheses predict consequences in accordance with specified relations among operationally defined variables. Why shouldn't we construct plans in the form of hypotheses so that we can rigorously test plans and the planning process.

If the Sackman approach is used it would be legitimate to use the rational planning model as a basis for setting goals and working towards them only if this model were perceived as a set of hypotheses being empirically tested throughout the development of the planning process. If it worked perfectly in one situation then the question would be asked, 'Can we identify the conditions associated with this situation which made this type of model so effective?' And the proposition made, 'Let us test it out in other situations where conditions are different, to establish if it is equally effective.' In this way a body of carefully tested planning theory could be developed which would by its nature be interdisciplinary because it takes account of technical, financial and human variables.

In Sackman's view this experimental approach would not be carried out solely by the elite groups in the situation – the systems designers and managers – but there would be involvement at every level of the user community. Ideas for testing would be formulated and modified through discussion between planners, managers, designers, users, operators and technicians. Sackman describes his approach as *systematic eclecticism*. Planning is eclectic in that it draws on all available ideas and its justification is pragmatic.

In accordance with the philosophy behind this experimental approach Sackman defines planning as,

> plastic evolving hypotheses concerning system objectives and system performance in specified environments, including embedding ecosystems to achieve operationally defined effectiveness levels, within stated resources, throughout the life cycle of the object system and successor systems.

This definition can be understood and simplified by setting it out in the following way.

- plans are defined as hypotheses
- planning is placed in an evolutionary system context
- plans must be operationally defined so that they can be tested out in the system setting
- the environment of the object system also includes the ecosystem in which it is embedded*
- plans are plastic human creations of desired futures within time and resource constraints
- plans are placed squarely in the middle of the real world
- the definition underscores the fallibility of the 'best laid plans of mice and men' by insisting on the need of accountability through continual testing in an uncertain world.

In this way Sackman presents us with a novel way of dealing with uncertainty which incorporates a very powerful learning process. Learning is achieved through testing planning approaches out empirically and observing which are successful and why they are successful.

This is an imaginative method which will lead to a systematic and increasing understanding of the planning process and the planning environment. However, its application requires a knowledge of scientific method and is inevitably slow. We believe that a great deal of information on the nature of the uncertainty in the planning environment exists already but is not known or understood by planners.

There is an increasing recognition of the fact that a perfectly rational planning process is unattainable because it requires certain conditions in the planning environment which do not exist in reality. Bauman (1967) sets out the requisites for perfect planning and indicates some of the reasons why it cannot be achieved. First, the planning system must be free from *external disturbance* so that it does not have to respond constantly to changes in its

---

* An ecosystem is a self organising biological system in which the organism interacts successfully with its natural environment.

environment which force it to redefine its goals. This implies that it should have total control over all the resources essential to the planning process, whether these are physical, or human. These resources should be under the control of the planning agent so that he can manipulate them as he wants. Clearly, few real world situations come anywhere near to this. In the case study firms described in the following chapters we found that the planning environment was constantly being disturbed by the appearance of new technological resources such as new items of computer equipment. These innovations presented the planning group with additional technological alternatives which had to be incorporated into the planning processes.

Within the firms themselves there was an unending battle between groups associated with the planning, systems design and implementation processes for control over scarce resources. Each group wished to manipulate these scarce resources (which were often intangibles such as status and influence) in terms of their own needs rather than the needs of the principal planning agent (which was usually the computer specialist group, or a section of this).

The perfect rational planning model assumes either that there is complete identity of interest between innovators, top management and user groups, or that a completely authoritarian system exists in which the planning group is able to eliminate all opposition so that negotiation and a reconciliation of con-flicting interests is unnecessary.

Second, Bauman states that the perfect planning model requires that all *information* relevant to planning activities be in the possession of the planning agent. This includes all information concerning the availability and possible use of resources as well as the technology of manipulating them. Failure to have all relevant information means that planning cannot be perfect because opportunities are missed through a lack of awareness that they can be attained. Firms attempt to maximise their use of available information through using experts, such as computer specialists, as principal planners and we note in our case studies that many of these groups spent the greater part of their time on extensive searches for information. When this was technical information their searches were effective and comprehensive and they had a good knowledge of the technical possibilities open to them at any one moment in time. But their lack of understanding of other aspects of the change process, such as the need for altering attitudes and values, and the desirability of changing organisa-tional structures, meant that searches for information in these areas, if carried out at all, were inadequate.

Third, Bauman suggests that the perfect planning model requires that the planning agent be capable of making a *decision* that is not only realistic but also most effective in terms of overall system goals. This in turn requires that the planning agent be,

- depersonalised, free of any motivations which are not identical with the pre-established goals of the system as a whole.
- competent in the sense of being able to choose among many alternatives

how to use available resources to achieve the best solution; this implies knowledge and skill and executive power.

Neither of these requirements were met in the behaviour of our decision makers. We shall see that much of the behaviour was influenced by the desire to achieve personal or subgroup goals which had little to do with overall system goals. Bauman also says,

- The alternatives among which selection is to be made must be reducable to a common denominator, commensurable, exhaustible by a simple and universal quantifying and quantifiable measure.
  To be subject to selection, alternatives must be compared; to be compared they must be comparable; to be comparable they must possess some common dimension.
  They should differ from each other only in terms of more or less in their quantities, not qualities.

This condition is particularly difficult to achieve when planning for computer systems. The firm in our sample which carried out the most exhaustive search processes in order to obtain comparative information to assist decision-making, found that in the hardware field alone different manufacturers were offering very different kinds of computer system. This made direct comparison of one system with another virtually impossible. A further problem when attempting to compare alternative systems on economic factors such as cost is that the economic evaluation of this form of technological innovation is still at a rudimentary stage. Certain cost factors such as hardware, software and the salaries of computer specialists can be derived without too much difficulty, but many of the major costs associated with the introduction of computer systems lie in the human area and are social rather than technical costs. These are extremely difficult to estimate in advance.

Fourth, perfect planning implies that the *social system* is homogeneous in the sense that there are no events which are simultaneously beneficial to one part of the system and harmful to another. In other words there must be no intergroup conflicts of interest. These bring into the planning process unpredictable elements which are difficult to manage and disturb attempts at rationality. In our case study firms we found that many different interests were associated with the planning, design and implementing of computer systems. Within the specialist group itself there could be different and conflicting interests between the programmers and systems analysts whose relative status and influence could be affected by changes in technology. Between the computer specialists and user departments there were often serious differences in the emphasis that was put on particular system goals, with the computer specialists focusing on technical goals and user department staff very concerned that their human needs and interests should not be lost sight of.

Lastly, the perfect planning process has no place for autonomous power and influence at lower levels in the planning mechanism. Perfect control of the planning process implies that all behaviour derives from the decision-making and executive power of the planning agency and its agents. Here is yet another uncertain area of the planning and decision-making process in a democratic society. In our firms we found that planning, design and implementation activities were complicated by attempts by groups who perceived themselves as excluded from significant aspects of the decision process to make more impact on what was taking place. Where decision processes were associated with a democratic approach and serious attempts were made to involve interested parties at every level in the decisions that were being taken, these problems were less serious.

One of the objectives of our study is to obtain a better understanding of the processes involved in changing an organisation from one sociotechnical state to another, particularly the nature of the internal environmental uncertainty with which the planning processes have to cope. In our research a great deal of this uncertainty could be identified as political in origin, for it stemmed from attempts by certain groups to use the planning process as a means for increasing personal power and influence and by other groups to preserve their existing power positions through maintaining the *status quo*. This form of uncertainty led to a great deal of bargaining and negotiation between the different groups and mechanisms for facilitating and containing this at either formal or informal levels had to be created within the planning procedures.

Planning is very much a political process, as the setting of new goals and the movement towards these new goals must disturb the status hierarchy of the organisation, yet the rational planning models of those theorists who set out principles for action do not incorporate this critical variable. Sackman's experimental approach would identify those aspects of planning most affected by political issues but would lead only slowly to a comprehensive theory of the influence of internal politics on the planning process. In this book we attempt to provide the theory and hope at a later date to be able to test it out systematically.

Change will always be associated with conflict and uncertainty; this is unavoidable. But it may be possible to ensure that unnecessary and additional conflict is not engendered simply because those associated with the change process do not understand the political and other factors which generate uncertainty.

Let us follow the advice of Ozbekhan and design the kind of future we want and plan to attain this, while recognising the uncertainty and complexity of the process we are embarking on and the powerful impact on it of human needs and values.

3

Decision theory and environmental uncertainty

In the last chapter we looked at the approach to planning most commonly found in the literature and the extent to which it recognised and analysed the uncertain environment in which planning takes place. In this chapter, we take one aspect of the planning process – the decision to make a change – and see how decision theory copes with the problem of uncertainty.

An interesting and important stage in the introduction of any innovation is the taking of a decision to make a change. As Simon (1961) points out in his book *Administrative Behaviour*, any practical business activity involves both 'deciding' and 'doing', thus an understanding of administrative theory and action requires a knowledge of the processes of decision which precede the actions taken.

Some of the questions we examine in this chapter are: How do firms make their innovative decisions? How do they identify problems that require solution? What search and choice processes do they carry out to determine appropriate courses of action? How do they establish that they have made the right decision? What problems of decision-making do they find most difficult and take most seriously, what problems do they find easiest to solve? Do their decision processes take account of all relevant environmental facts likely to affect the success of the decision, or are they only aware, or prepared to take account of, some factors and not others? In subsequent chapters we try to answer these questions by analysing the behaviour of our case study firms.

First we need to clarify our ideas on what is meant by 'decision-making'. We need to examine different kinds of decisions and specify those which are relevant to our study. We need also to consider theoretical ideas on how decisions are and should be made, together with practical problems experienced by managers in a decision-making role, and environmental factors located both within and outside the firm which influence the manner in which decisions are made and the kinds of decisions managers choose to make.

Some theories and ideas on decision-making

What is decision-making?

Let us begin by stating how we are proposing to use the term 'decision-making'. Simon (1965) defines it as 'becoming alert to a problem, exploring it and analysing the different components of the problem and finally deciding on a course of action'. For Simon decision-making is synonymous with managing. Etzioni (1968) takes a somewhat narrower definition and defines it as 'making a conscious choice between two or more alternatives' and selecting the most appropriate means to achieve an end. He believes that the range of options available to be chosen between is normally much greater than the decision maker is aware. For the purposes of the analysis presented in this chapter we take Simon's definition but even this greatly oversimplifies the complexities of the kinds of decision processes with which we shall be concerned.

Some varieties of decision-making

Simon (1965) has made one of the clearest and simplest categorisations of decisions. He divides decisions into two polar types, programmed and non-programmed, with a continuum in between. Programmed decisions are repetitive and routine, decisions which a firm is accustomed to take at regular intervals and for which it will usually have developed some form of systematic procedure. Many people would argue that these are not true decisions because little uncertainty or risk is involved. They might be more properly called logical deductions. Non-programmed decisions, in contrast, are novel, unstructured and usually very significant. Here management is called upon to respond to very uncertain decision situations where no procedures for choice exist and where it has to rely on its own capacity for intelligent, adaptive, problem-orientated action. Simon suggests that man is seldom completely helpless in a new situation, he has a problem-solving capacity which he can use when faced with something which he has not encountered before. Problems requiring decisions which fall initially into Simon's non-programmed category will of necessity become programmed if they recur a number of times within a relatively stable environment. Most managers and organisations will attempt to order and accelerate their solution by developing systematic procedures for handling them. Technical decisions associated with the use of computers were non-programmed in the early days of the technology and this is still true of new and novel applications. But firms which have been using computers in commercial areas for many years are now starting to develop procedures for identifying priority areas for computer usage, and for deciding which applications will provide the greatest return on outlay.

Simon describes the traditional techniques available for assisting the manager to make both programmed and non-programmed decisions. For example, habit has been the commonest way of making programmed decisions;

we approach a problem in a certain way because we have always approached it in that way. Then, many managers and firms have developed standard operating procedures. These are formal, written programmes used for in-doctrinating newcomers into the decision-making norms and habits of the organisation, and for reminding old hands of the commonly used methods for solving particular problems. Where programmes for solving problems are not written down then organisations may secure uniform behaviour patterns through developing common goals, norms and expectations in their staff. These ensure that staff will tend to behave in a similar way when presented with customary situations. There have been far fewer techniques available in the non-programmed area. Simon talks of the use of judgment, rules of thumb and the selection of executives with good problem-solving skills – generally a capacity for logical thought. Attempts are then made to develop these somewhat nebulous attributes through training and experience.

A most important form of non-programmed decision and the one with which we are concerned in this book is the innovative decision. This is de-fined by Knight (1967) as 'the adoption of a change which is new to an organisation and to the relevant environment'. Such decisions contain major elements of uncertainty. From the firm's point of view there will be uncer-tainty over whether the introduction of the proposed innovation will pay off in financial or efficiency terms. From the point of view of any specialist group associated with the introduction of the innovation there will be uncertainty over their competence to create a new organisational form. From the point of view of the line manager who has to use the proposed innovation there will be uncertainty over his competence to operate it successfully, and its benefits for his personal goals and the goals of his department.

Knight sees innovative decisions as a response to two kinds of situation. There are those which occur when the organisation is not under pressure. Things are going well, there is a degree of slack and the firm begins to search for new ideas, new products, technologies, and processes in order to maintain and increase its competitive position. In contrast innovation can be a response to a distress situation. The firm seeks change in order to survive; but it is now more likely to seek improvement through the cleaning up of existing pro-cedures than through the more risky approach of bringing in new products or technologies. The introduction or extension of computer usage can be a response to either kind of situation. With the first, EDP is introduced in order to secure something thought to be desirable but unobtainable with existing technology; for example, management information for use in de-veloping corporate strategy. With the second, EDP is likely to become a vehicle for tightening up internal controls so that management can keep a firmer hold on in-company activities.

Decision-making as a part of planning and policy making

Decision-making has sometimes been studied by academics as if it were the prerogative of individuals or small groups presented with micro problems. In reality it is frequently a macro process carried out by institutions and societies and some of the most important studies of decision-making have been made by students of national planning, the work of Gross (1967) being noteworthy here. Even within a single firm decision-making has to be considered within the framework of an organisation responding to its environment by developing policies for innovation and then implementing these policies through the design of plans for action. This kind of process will involve the making of a very large number of decisions and decision-making will be shared by a number of different groups within the organisation.

Increasingly, in the future, we can expect pressure for the democratisation of planning, and these groups will be located at every level in the organisational hierarchy. One of the most interesting areas of investigation for the student of innovative decision-making is an identification of who is participating in particular decision processes and the nature of the different roles played by each group or individual. Innovative decision-making in large organisations is essentially an interactive process in which many different interests have to be reconciled.

Theoretical approaches to decision-making

Simon (1965) follows the rational approach described in the last chapter and sees decision-making as covering *intelligence* activity, *design* activity and *choice* activity. Intelligence activity involves becoming aware of a problem and the possibility of solving it. Design activity is the process of testing out alternative means of solution, and choice activity is deciding which of these alternatives will best solve the problem in a way which fits with the objectives of the organisation. Most decision theorists would accept that these activities *do* take place but there are many different opinions concerning the manner in which they take place.

Cyert, Dill and March (1967) draw our attention to the fact that the two best known theories of decision-making look at the process from very different points of view. The first is derived from economics and treats business behaviour as a rational attempt to maximise profits, with organisations knowing whether or not they are achieving this goal and knowing also the courses of action necessary to enable them to do so. Business is perceived as a highly logical activity in which the organisation constantly scans all alternative means for achieving its profit objective and adjusts its investment portfolio accordingly. Traditional economic-mathematical theorists view decision-making as a rational choice process in which an individual or an organisation is presented with a number of alternative courses of action. Each alternative has a set of consequences and the individual has himself a set of preferences. These preferences enable him to rank consequences and to choose the alternative that has the preferred consequence in terms of his objective which, in the case of a business decision, will be the maximisation of profit. This theoretical approach implies that firms have accurate information on the costs to be incurred and the benefits attainable through adopting particular courses of action, and that decisions are made on the basis of this information.

The second major theory of decision-making has a more behavioural stance. This school, of which the best known exponents are Cyert, March and Simon, sees human decision-making behaviour in organisations as encompassing searching, choosing and problem-solving, but with these activities subject to human restrictions such as the limited amount of information that a human being can handle, and the limited ability of a man to do more than a few things at one and the same time (Cyert *et al.*, 1956, March, ed., 1965; March and Simon 1966; Simon, 1961, 1965). Because of these human shortcomings man does not usually seek for optimal solutions, but accepts solutions which solve a particular problem satisfactorily although not necessarily in the best available way. This restricted behaviour is known as 'satisficing' and is in contrast to the 'optimising' approach of the economists. This theory appears closer to reality than that of the first school, though we show later that the decision behaviour of our case study firms was subject to even more complex influences. For example, they accepted satisfactory solutions not merely because of their inability to identify optimal solutions but because of the variety of objectives the innovative decision was seeking to meet, and the network of different interest groups associated with the decision process. An optimal decision would have meant that there was one clear objective for computer innovation which all groups accepted and were prepared to realise, and we did not find this in any firm. Also these were risk decisions involving a large element of uncertainty and a high degree of capital investment; standard operating procedures for dealing with this kind of decision did not exist, so the search process and evaluation of alternative solutions was extremely exhaustive and subject to many different kinds of pressures.

Etzioni (1968) has criticised both these theories of decision-making and modified them to fit his own behavioural view of decision activity. He points out that the rational economic-mathematical approach precludes the examination of what he calls 'irrelevant' considerations – pressures to choose alternatives which are not related to the best way of achieving a particular goal. For example, selecting a solution because it will 'please the boss' or enhance the decision maker's own power and influence in the organisation. These are 'value' considerations based on an individual's culturally defined perceptions of what is a desirable role in his firm and how he can achieve this through his decision-making activities. Etzioni is here expressing a view which is becoming common among behavioural scientists, that decision-making can only be understood if it takes account of the values of the society in which the decisions are being made. Etzioni goes even further than this suggesting that the 'detached' man is not necessarily the most effective man and that decision makers should constantly incorporate value considerations in their decision criteria.

With innovative decisions it may be essential to be aware of those solutions which will prove personally acceptable to the managers who have to operate the innovations. Etzioni also suggests that decision-making will not be effectively carried out unless it provides some rewards for the decision maker. If he has little interest in either the processes of decision-making or the final goal, then his problem-solving is likely to be poor. With macro decisions, of which the decision to introduce a new computer system is a good example, decision makers will have to cope with a multitude of different goals and different values at one and the same time. They therefore have to develop skills which help them to become aware of key issues at an early stage and create procedures which enable them to order the great variety of factors they have to investigate.

Etzioni's principal criticism of the traditional approaches to decision-making is similar to that of Simon. This is that they are not and, in fact, cannot be used. Securing information about alternative courses of action is extremely difficult as alternatives keep changing; this is especially true in the computer field where new technical possibilities are constantly being provided by manufacturers. A calculation of the consequences of each alternative is also difficult if not impossible to achieve as many of the consequences of a new technology will not be known and they may be extremely difficult to predict. Also agreed values on goals may hardly exist. The innovator's goals may differ from those of top management, while the goals of the line manager who will use the new system may be greatly at variance with those of the innovator. Simon's solution as we have seen is to 'satisfice' or to be as rational as you can. Administrative man has to construct a simplified model of the real situation in order to be able to cope with it and he behaves rationally only in respect of his limited personal model.

Etzioni puts forward another theory of decision-making which can be termed the art of 'muddling through'. With this approach large-scale decisions to change things are not taken but small reforms are constantly made in an incremental manner. Etzioni calls this 'incrementalism'. If the decision-making area is a new and uncertain one, as may be the case with a new technology, then a low level decision-making model is required as it becomes extremely important for a management not to risk complete disaster through too large a step into the unknown. Muddling through involves only the investigation of those alternatives that do not differ too much from existing practice and do not require a preliminary consensus on goals from a number of different groups. Consensus may be restricted to the decision-making group and ends are chosen which are appropriate to the means available. With this approach problems are not solved but merely attacked and the primary objective may be 'improvement', 'doing better', rather than some large, specific goal. Bauer (1966) sees muddling through as a peculiarly British approach. He suggests:

> The doctrine of muddling through is based on elaborate concern for second-order consequences of actions. It assumes that social systems and processes are very complex phenomena and that it is impossible to determine in

advance exactly what results will be created by one's actions or what difficulties will be encountered.

The difficulty with this approach is twofold. First, as Etzioni points out, decisions arrived at in this way may over-represent the strong – those involved in the decision-making processes – and under-represent the weak and uninvolved. The fact that consensus is only required of the formal decision-making group may mean that changes are made that are unacceptable to those on the receiving end. Bauer (1966) sees the means for overcoming this possible defect as responsiveness to feedback: 'The doctrine of muddling through has a contemporary look in its sensitivity to feedback from the environment, and its disposition to change tactics when the data fed back suggest that the results produced differ from those intended.' A second and more serious difficulty is that this approach may not be appropriate for new forms of technology such as computers. Here a decision to introduce a new kind of hardware may require major changes, the consequences of which are difficult to predict, and a policy of incrementalism may prove impossible.

Etzioni's (1968) preferred solution to the decision-making problem is what he calls 'mixed scanning'. This is a combination of the rational and incremental approaches in which fundamental decisions are separated from small decisions. The former are based on the rational model as modified by the Simon school, the latter on the incremental model. Fundamental decisions are treated to a broad search process which is high on the coverage of alternatives but low on detail. Minor decisions relating to subunits of the larger problem are given less coverage of alternatives but are much higher on detail. This means the approach to major decisions is to list all relevant alternatives; to reject those for which there is some major objection; to re-examine the remaining alternatives and reject those to which there is some objection, and to continue with this process until there is only one alternative left. Minor decisions are not treated to this evaluation and rejection procedure: a few alternatives only are considered but great attention is paid to assessing consequences, with the intention of constantly improving any change that is made.

This approach of Etzioni's is interesting, but it is difficult to see how it greatly improves on existing theories of decision-making, nor does it appear to give any very useful guidance to those concerned with making large-scale innovative decisions. Here the problem may be of finding a basis for comparing alternatives and selecting valid criteria for accepting some and not others. The theories of decision-making described so far provide very little guidance in either of these areas.

All the theories we have considered appear to have serious limitations. Simon very rightly criticised the earlier rational economic mathematical approach for assuming that a decision maker would know all the alternatives open to him. He brought the theory of decision-making a step further forward through introducing the concept of 'search' and the need for the decision maker actively to seek out those alternatives that he should consider. He also drew attention to the fact that the realities of decision-making led to the acceptance of 'satisficing' solutions (solutions which would satisfactorily solve a particular problem) rather than optimising solutions (the best solution to a particular problem). Most students of decision-making now agree that the comparison of alternatives is not usually made in terms of a single criterion such as profit (see Cyert *et al.*, 1956). With innovative decisions in particular there are generally important consequences which are so intangible as to make an evaluation in terms of profit difficult or impossible. Again problems which require decisions about solutions are not always instantly recognisable. In the modern firm with its specialist groups such as management services and R and D, an important activity is the searching out of problems which require organisational attention. Our case studies have shown that groups such as management services departments have a vested interest in change. The logic of their existence requires this. One of their most important roles is the identification of organisational problems to which a computer based solution can be applied.

Braybrooke and Lindblom (1963) have also provided a number of criticisms of existing theories of decision-making. They point out the little account that is taken of the closeness of the relationship between facts and values when decisions are being made, the absence of a satisfactory evaluative method, the openness of the system variables with which decision-making contends, the decision maker's need for strategic sequences of analytical moves and the diverse forms in which policy problems actually arise. Etzioni (1968) also stresses the fact that 'universal' theories of decision-making have serious shortcomings as they do not consider the forces generated in the culture in which the decision has to be taken. Despite these limitations, the majority of the theories we have examined do recognise environmental uncertainty as a factor in decision-making, and those of Braybrooke and Lindblom and Etzioni identify human value systems as an important factor generating uncertainty.

Practical aids

In recent years a great deal of attention has been paid to methods for improving a manager's decision-making skills. Operational researchers have developed mathematical models to assist decision-making and sociologists and psychologists have tried to draw management's attention to significant behavioural variables of which decision makers should take account. Managers with responsibility for repetitive routine decisions, the kind Simon calls programmed, have benefited greatly from the development of such aids. Many firms now have standard operating procedures, documented in considerable detail, to which a manager can refer when faced with a particular kind of decision. Generally these take an algorithmic form. A precise set of steps is set out which will lead to the required solution. Innovative decisions of a non-programmed kind can be assisted by heuristic approaches. A heuristic is a process for solving a problem which may aid in its solution, but offers no guarantee of doing so. Taylor (1965) describes a number of these. For example, *working backwards*, or beginning with the result you wish to achieve and working backwards step by step to that which is given; looking for an *analogy* or comparing the current situation with some problem you have settled successfully in the past; *making a plan* or finding a problem similar to the one you are trying to solve, but simpler: solve the simpler problem and use the same approach for solving the more difficult problem.

Operational research also provides some specific tools for the solution of non-programmed managerial problems; for example, linear programming, dynamic programming, game theory and probability theory. These can be used for constructing mathematical models which mirror the important factors in the management situation to be analysed and the manipulation of such models can provide a manager with much useful information on which to base a decision (Simon, 1961). The danger with this approach is that of oversimplifying complex problems in order to make them amenable to mathematical analysis, for in doing this all contact may be lost with the realities of the problem and of the environment in which it is occurring.

Unfortunately many problems handled at middle and top management level are not amenable to mathematical treatment and Simon (1961) suggests that for this type of problem we need two things. First an understanding of how to increase the problem-solving capacities of people faced with non-programmed decisions and second, greater knowledge of how to use computers as aids to human decision-making without having first to reduce the problem to mathematical terms. Both of these require a greater understanding of human decision processes and of the decision environment than we have at present. In real world situations it seems that attempts at what is seen as rational decision-making frequently become distorted and overlaid by political issues which are not always recognised or made explicit. When this happens no decision-making aid is likely to be of much use for decisions will be taken on the basis of a weighing up of the political situation rather than a choice made in terms of organisational goals. Despite all the problems described above Simon rightly points out that decisions *are* made, and that the effectiveness of the results achieved is often out of all proportion to the groping, almost random processes we observe when problems are being solved.

Problem definition and goal setting

The first stage in any innovative decision is a recognition that the firm has a problem which is worth investigating to establish if it should be solved. K. E. Knight (1967) suggests that this problem recognition is a product of an organisation's inability to achieve the goals it has set, or of a perception that its ability to do so may become less in the future. Assessment of goal achievement is likely to be based on historical comparisons of how well the firm has done in the past rather than on some absolute level of performance (Cyert, Dill and March, 1967). Moreover because different subunits within a firm have different objectives there are likely to be different opinions of what constitutes success. Feelings of failure as a stimulus to innovation is a rather negative response, however, and as we have seen may lead to an improvement of existing ways of doing things rather than to a revolutionary form of change. A more positive approach is when an organisation sets itself new goals and innovates to achieve these. Many firms today use their specialist groups constantly to scan the organisational environment, with the objective of assessing when and how the firm needs to change its goals in order to adapt more easily to environmental pressures such as new product market demands or new technologies.

Firms can become better at spotting existing or potential problem areas by organising themselves to facilitate the problem identification process. There are many ways of doing this. Firms may recruit or encourage what Leavitt (1958) calls the maverick executive. The individual who is attracted by the new and is prepared to take personal risks in trying out new products and processes. We found a number of such people in the management services departments of our case study firms. One had a sign over his desk which read 'if it works it is obsolescent', and viewed his own group of specialists as an insubordinate minority within the company whose major function was to prove to top management that existing ways of doing things were basically wrong. Another approach is to enlarge jobs so that people are given responsibility for identifying problems. Brown (1960) has always stressed the importance of giving managers developmental responsibilities so that they are constantly seeking to extend and improve their role. Morris (1971) sees encouragement of this developmental aspect as an important part of management development programmes. Lastly, firms may encourage innovative search mechanisms through the creation or encouragement of R and D groups whose brief is the seeking out of new ideas. Leavitt (1958) points out that many firms are worried that their research groups do not make a sufficient contribution to company profits. In his view they should be more concerned whether their research groups are directing too much attention to routine programmed research activities instead of concentrating their attention on new ideas.

A difficulty for managers involved in innovation is not merely spotting that a problem exists, but asking the right questions about it. For example: Can we specify what the problem is with precision? Should we do anything about it? If yes, when and how? Can we put boundaries around the problem in order to limit the area of innovation? How much will it be worth our while to spend on the solution of the problem? These, and many other similar questions, need to be asked and answered before the firm takes a decision to proceed with any innovation. Churchman (1968) suggests that the essence of decision-making is asking the right questions, not simply providing definitive answers.

The decision to proceed with innovation

The first decision about any problem is whether to continue to live with it or whether to set in hand investigations on how it might be solved. Rogers (1962) suggests that the probability of going ahead depends on the nature of the innovation envisaged. Innovation will take place if the likely solution has obviously great advantages over current activities, if it is compatible with current values and experiences, if it is simple enough so that no extensive training or extraordinary efforts are required, if it is easily divisible so that various parts can be tried out, and if it is easily communicated so that its consequences are readily apparent to both the innovator and his reference group. Our experience suggests that these conditions do not hold good for technological innovation such as the introduction of new computer applications. The first is certainly true – a belief that a computer based solution to a problem has great advantage over current methods, but our firms were keen to go ahead with this form of innovation even though the anticipated consequences of computer technology were not always compatible with current values and experiences. This incompatibility might be particularly true of the user area where considerable training and new attitudes to work would be required for successful implementation, and where the consequences of a new system might be unclear and difficult to demonstrate in advance to interested parts of the firm. We came to believe that the attraction of new forms of technology, the lure of the 'technical ethic', was more powerful than considerations which might inhibit the introduction of non-technical kinds of innovation.

The student of organisational behaviour is interested in establishing which groups first identify problems requiring solution and which groups take the decision to proceed with an investigation. These are not necessarily the same people and with computer innovation it is frequently the management services specialists who delineate the problem and top management who take the decision to proceed further. In our experience the approach of management services to top management frequently takes a 'selling' form. That is, management services stress the potential seriousness of not tackling a problem and persuade top management that a computer based solution can provide the most effective remedy. Investigations are then carried out by members of the management services department who feed information for further decision-making upwards to top management.

A survey described by Reddington (1972) found that many heads of management services departments secured agreement to computer innovation in this way. A wise top management may wish to introduce some checks at this point to ensure that it is getting all the relevant information from its management services department and that the information it receives is not unduly biased towards a computer based solution to the exclusion of other kinds of solution. At this stage it may be greatly to top management's advantage to decide not to go ahead with a proposed innovation if the likely benefits cannot be clearly demonstrated. Experience shows that as money and time is spent on long-drawn-out investigations so it becomes increasingly difficult for a management to stop the process and pull out. Barnard (1938) points out that the art of decision-making lies as much in deciding not to do something as in deciding to go ahead. Top management may feel that it does not have enough information, or that it is too early to innovate in this area, or that a proposed innovation would not be operationally effective because people would not use it, or that the costs would be too high and the benefits too intangible.

The setting of goals for innovation

A decision embodies a purpose to do something for a number of what should be clearly thought out reasons. The decision-making group has to be able to sift out the facts which are immaterial and irrelevant to the decision, and to distinguish those facts that assist the accomplishment of the purpose from those likely to hinder its accomplishment. Most important, however, is being entirely clear about the purpose the organisation wishes to achieve, and this implies setting realistic goals and objectives which can be achieved by means of the innovation. This process is complicated by the fact that organisations will frequently be using an innovation to achieve a diversity of goals. Broad goals will need to be set as soon as the decision to proceed further with the investigation is taken, otherwise the search for information will be too wide ranging and nonspecific. The decision group will need to recognise that as the search process throws up information so early goals may have to be changed to take advantage of unanticipated constraints and opportunities. Also, all organisations today operate in rapidly changing external environments and goals are likely to have to be adjusted to meet new pressures and demands which originate in the product market which the firm serves. It is unlikely that many contemporary organisations will operate in stable enough situations to enable the same set of goals to continue unadjusted throughout a major change process.

The formulation of goals requires a logical decision-making process at top management level, and one that is related to the overall corporate strategy of the firm. Only if innovatory goals are clearly formulated can management start to think through the means by which these can be achieved. In the past a danger associated with the use of computers has been an attraction to this technology as a means for change without sufficiently precise thought about the kinds of changes likely to prove most worth while. The result has been a proliferation of computer systems that have not been regarded as financially or administratively successful.

A problem which technological innovation shares with other forms of change is obtaining an adherence to organisational goals on the parts of all groups concerned with the change. Group and individual interests may mean that departmental or personal subgoals are given priority over the major goals the organisation is seeking to achieve. The influence of these kinds of subgoals on decision-making seems often to be unrecognised by decision theorists who do not appreciate a point made by Soelberg (1963) that, 'desire for power and concern for personal advancement represent goals which are of central concern to an organisational theory of decision making'. Our experience of the introduction of computer systems suggests that all personal goals that affect decision-making are not necessarily of this aggressive kind. Managers often seek a quiet life, or a situation where they may undertake work that is familiar to them and which they are good at. The decisions they take and their reaction to innovation may be just as much a product of this kind of goal as of the more aspiring kind. Nevertheless, whether a desire for power or a desire for security influences attitudes to change, people who feel threatened increase the uncertainty of the decision-making situation.

At the early stages of the innovation process 'the planning group' will be most interested in setting strategic goals and objectives. These will broadly indicate the kind of situation management hope to achieve as a result of introducing the innovation. This may be a more efficient accountancy system, a production department that can realise more precisely its production targets or a management that can respond faster and more efficiently to problems because of the introduction of a new management information system. Other kinds of goals will become necessary as the decision-making process progresses. These will be tactical goals relating to the means for introducing the innovation and may include operational or political goals directed at damping down any attempts at opposition. In a later section of this book we set out some of our observations on how political factors can have a major effect on an innovative decision process.

The search for alternative solutions

Many new computer systems will come into a 'large project' category. Wiest (1967) defines a large project as

> an extensive, one-of-a-kind undertaking leading to a final, well defined goal or product. Projects typically consist of several hundred or thousand separate activities or jobs, at least some of which must be done in a given sequence. . . . Each job normally requires one or more resources (men, machines, money) and a given time for completion.

Any search process which is carried out to identify alternative solutions for a large project will cover a number of different subproject areas, some of which may be sequential in that they require an early problem to be solved and a decision taken before later problems can be tackled.

The importance of search behaviour as a factor in decision-making was first described in detail by March and Simon (1966). They argued that information is not given to the firm but has to be obtained; that alternatives are searched for and discovered sequentially, and that the order in which the environment is searched determines to a substantial extent the decisions which will be made. All this is true, but the search process appears to be even more complex than that described by March and Simon. Our observations show that search is often a two-way or 'mating' process. A firm is seeking for solutions while at the same time solutions located in the firm's environment are looking for problems. This is particularly true of computer innovation where a band of highly trained computer manufacturers' salesmen are constantly on the lookout for firms with problems which they can help solve through the technology they are marketing. An additional uncertainty is the fact that alternatives are not fixed but keep changing. Computer technology is developing rapidly and each successive development can offer a user firm new means for solving its problems. Management may have decided to solve a problem in a particular way, when a new piece of computer hardware appears on the market offering such apparent technical advantages that the firm is persuaded to review the decision it has already made.

Again, solutions in major projects usually involve the reconciliation of different sets of values and interests, and this is a factor that has to be taken account of in the search process. It may be fruitless to pursue certain lines of investigation simply because a solution of that kind would prove unacceptable to the people required to operate the new system. Another difficulty for the decision maker in his search for alternatives is the fact that often he has no way of knowing when he has found the solution most likely to achieve the goals the firm has set. He may not have legitimate criteria for making comparisons between alternatives or, once the new system is in, of testing that the correct choice of solution has been made.

In addition to these uncertainty generating factors, investigators and decision makers are subject to another set of pressures during the search process. They will have to reconcile their personal goals and the goals of their departments with the strategic goals of the organisation. They may also have to take account of political and psychological factors which influence opinion at the top of the company. Senior management has its own personal idiosyncrasies, needs and aspirations and solutions may have to be rejected if they are known to be unacceptable to the interests of powerful members of this group. Other problems relate to how widely to extend the search and when to stop it. Searching for solutions is in itself a costly and time consuming process and the costs of undertaking it represent real financial factors which need to be incorporated into any final estimate of costs and benefits.

The shape of the search process

Clearly any meaningful search process is not going to be easy and will have to cover many different stages of project design and implementation. Generally search processes will have a shape, that is they will extend sideways as alternative solutions for different stages of project design are investigated and they will extend forwards as the solving of early problems leads to a concentration on later problems. There will also be a looping process as later solutions force a rethinking of earlier solutions. The process is set out in Fig. 3.1.

We cannot agree with March and Simon (1966) that alternatives are searched for sequentially. With large projects it is quite usual to split an investigating team up into small groups and to give each group simultaneous responsibility for examining a different alternative. We accept March and Simon's view that investigators do not search out all alternative solutions available to them but seek a satisfying rather than an optimising solution. To search out all solutions in a technical area subject to constant change and development would clearly be a never-ending process. Later on we examine what factors cause a search process to be limited or halted. Simon suggests that the conspicuousness of alternatives is a factor in their consideration, also dissatisfaction with available solutions may be a stimulus to further search.

Fig. 3.1 Problem-solving: the search process

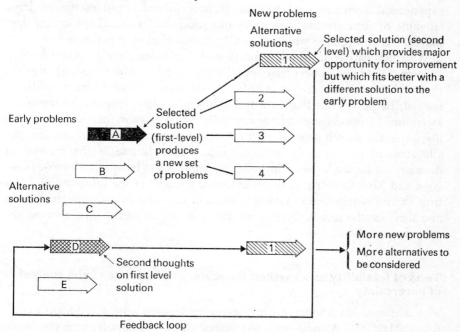

Simon (1956) also suggests that searchers become adept at recognising clues which suggest avenues of investigation that are fruitful to pursue further. To some extent this is a learning process as searchers will come to appreciate that the following of a certain kind of clue or stimulus tends to lead nowhere whereas the pursuing of a different kind generally leads to worthwhile information. Another limiting factor in the search process is the preliminary goals which the organisation sees itself achieving through the proposed innovation. It is unlikely that alternatives will be pursued if they are seen as leading to a quite different kind of goal from that sought. In our research we found that a reformulation of goals by the innovator group and by top management was a major factor in prolonging the search process and adding uncertainty to the planning environment, for each time goals were redefined new alternatives were examined.

One method a planning group uses to reduce and control uncertainty is described by Knight (1967). He points out that an individual or group will work within its cognitive limits when searching for a viable solution. That is, it will tend to restrict the search process to areas within its own experience and knowledge. Individuals and groups can only handle a limited amount of information at any one time and a major part of any decision-making activity is the reduction of masses of information to an acceptable level. It is in this process of variety reduction that management techniques such as operational research can prove of value. Handling variety may be as much a matter of practice as of cognitive limitation, however, and innovative groups who are experienced searchers and who know how to order and categorise large amounts of data are likely to carry out much wider search processes than individuals and groups which do not have this kind of experience.

With innovative decisions a fundamental question may be when to stop collecting data and concentrate on taking a decision. Morris (1964) suggests that a theory of the cost and value of information for decisions is badly needed; he points out that in a technological society, information costs are ascending to previously unimagined heights. It is now increasingly recognised that the search process itself has major costs attached to it and that the allocation of resources for securing information is one of the investment decisions to be made when introducing innovation (Koopman, 1956). Hawgood and Mumford (1971) have developed a model of the information that a firm should obtain before taking a decision to invest in a computer system, and they see the search for this information as part of the total costs of the project.

Tests of feasibility as a method for variety reduction and the control of uncertainty

The management literature contains many techniques for reducing uncertainty. McKinsey Associates (1968) suggest that new computer projects should be checked against three tests of feasibility; first, *technical* feasibility – can we do it? second, *economic* feasibility – is it worth doing? third, *operational* feasibility – will it be used? Technical feasibility is a matter of expert judgment and asks the question, 'Is this application possible within the limits of available technology and our own resources?' Economic feasibility is concerned to establish that the application will result in more financial benefits than it costs to develop. Operational feasibility checks that if the system is successfully developed it will also be successfully used – that managers will adapt to the system and not resist or ignore it. McKinsey *et al.* believe that asking these questions can provide the first information screening process for the decision maker. If the answer to any of them is no, then his search process need go no further and if he is wise he will turn his attention to another method for solving his problem.

McKinsey *et al.* point out that it is dangerously easy to avoid confronting the full implications of feasibility until the project is well under way. If the answer to all of the tests of feasibility is 'yes' then management can start considering technical, economic and operational alternatives and the detailed search process begins. In whatever way he approaches the decision-making problem the manager is likely to adopt a satisficing approach, not because of the difficulty of getting sufficient information to optimise but because any major innovation requires the reconciliation of a large number of different interests. Therefore a solution must be found that satisfies all of these at least to some extent.

Many managements adopt an even simpler approach than that recommended by McKinsey Associates for arriving at a satisficing solution. Cyert, Dill and March (1967) describe one firm introducing a safety innovation where the management did not concern itself with any comparison of alternative solutions but merely asked a question of the first solution that presented itself, 'Is it feasible?' Feasibility implied two things; first, budgetary constraints, 'Is the money available?' Second, an improvement criterion, 'Is the proposal better than existing procedures?' The reason for this simplistic approach was the difficulty of developing a single criterion on which all relevant considerations could be measured. Cyert and his colleagues rightly point out that the question, 'Is this change feasible?' is much easier to answer than the economic question, 'Will the return on this investment equal or exceed the return on alternative investments?'

The direction of the search process

We have seen that the shape of the search process is sideways and forwards; this shape appearing as more alternatives are considered and as one stage of the problem is solved and there is a move to the next stage. Let us now examine more closely the forward process. Clearly, if the problem is to be solved at all there has to be a definite progression towards a solution, but our experience when studying this kind of decision-making indicates that the search process moves through a number of levels with a major decision being taken at each level, each decision being followed by an examination of the next set of problems and a search for alternative ways of solving them. Frequently this new set of problems is a product of the decision just taken. But the progression from stage to stage is not always straightforward and a looping process is likely to take place as decision makers become aware of a new possibility, and this causes decisions made at an earlier stage to be revised. This looping is specially apparent in decisions made about the use of technology, for new technical possibilities are likely to be drawn to management's attention before the problem-solving process is complete.

Other factors are also likely to alter the direction of the decision process. With large strategic decisions managements tend to change their original goals as the problem-solving process progresses and they become aware of other problems and possibilities which they had not previously considered.

Again, decisions may be altered because of a reassessment of means when a management comes to realise that it does not have the resources to achieve the goals it first formulated. The entire decision-making process is therefore highly uncertain and dynamic, and an anxiety for managers taking decisions in the computer area is their knowledge, when the final solution is selected, that the new system will be obsolete as soon as it is introduced, for by that time new technological alternatives will be available. Stafford Beer (1972) describes this process as *absolutum, obsoletum*, if it works it's out of date. Thus, as McKinsey *et al.* (1968) point out, throughout any project there is a need for a continuous reassessment of the technical and economic risks and payoff probabilities in order to keep the project on the right track.

Some examples of approaches to the search process

The literature on decision-making still contains relatively few detailed examples of real life practice, but the best of these have come from Cyert, March and Simon and their research associates. Cyert, Dill and March (1967) provide a case study which describes the decisions taken after an overhead crane accident and a recognition by the firm concerned that accelerated reno- vation of old equipment was required. Their examination of the search be- haviour of the firm showed that this was severely restricted, and distinguished by local rather than general scanning procedures. The firm regarded its re- sources as fixed and the economic feasibility test imposed was 'Can we afford it?' and not 'What will give us the best return on capital invested?' Cost and return estimates were vague and readily changed and this fact made the de- cision susceptible to the ideas of the person paying most attention to it at any one moment in time. Top management was seen by those lower down as using the decision situation to promote favoured projects which it had not previously had an opportunity for implementing. Cyert, Dill and March concluded that once an alternative appeared it was accepted if it satisfied the general cost and return constraints and enjoyed the support of key people. Early scanning produced only a few alternatives and a firm commitment to action was taken before the search proceeded very far. Cyert and his colleagues suggest that this happened because

> Organisational search consists in a large part in evoking from various parts of the organisation considerations that are important to the individual sub- units, and the relevance of such considerations, and the impetus to insist on them, are not manifest until the implications of the decisions are made specific through innovation.

What they seem to be saying here is that a solution is seen as acceptable because its consequences are not completely known when it is taken, and these only become apparent once the solution is implemented.

Cyert, Simon and Trow (1956) have provided an extremely useful descrip- tion of a decision made by a firm in the early 1950s to use data-processing equipment. This was a more far reaching decision than the one described above and involved more comprehensive search procedures. They describe the search process as being concerned with a whole series of 'nested' problems. Each alternative solution at one level leading to a new set of problems at the next level. Also, the search process took a number of different forms. There was a search for *procedural* alternatives – different ways of using the same data-processing equipment and for *substantive* alternatives – different equipment possibilities. They found that these information gathering activities took up the most time in the decision process, and that the criteria used for decision-making were not necessarily the same for each alternative. Sometimes a choice was made on monetary criteria, at others on political criteria. This case study left a number of interesting questions unanswered concerning how organisations cope with uncertainty in the decision-making process. For example, what factors determine how many alternatives a firm considers before a decision is made? Under what conditions will an alternative tentatively selected be submitted to a further check? What type of search strategies are used to spot alternatives? To what extent do tangibles as compared with intangibles figure in decision-making? We hope that our own case studies will throw some light on all these questions.

The search for consequences

So far we have concentrated our attention on the search process in decision-making as it is concerned with coping with uncertainty by identifying alternative ways for solving a problem. A second search process which stems from the first is the search for the *consequences* of particular alternatives. The viability of an alternative cannot be established unless its consequences for the goals of the organisation, the goals of subgroups and the goals of individuals can be identified. Writers on decision theory appear to have little to say about this part of the decision process. The case study provided by Cyert, Simon and Trow (1956), although comprehensive in many respects, provides sparse information about the kinds of consequences which the firm examined or the extent to which it examined consequences at all. Other writings by Simon suggest that the search for consequences is carried out after all alternatives have been identified (Simon, 1961). Our experience is that this is not the case and that a rough screening out of alternatives on the basis of grossly undesirable consequences takes place almost as soon as an alternative is identified. A detailed examination of consequences takes place at a later stage in the decision process when the decision-making group has settled on a small number of apparently viable alternatives. Even when a search for consequences does take place this may be restricted to technical or economic effects. We have found that a weak area with computer innovation is the omission of an early identification of the human consequences of alternative courses of action. Often little thought is given to these and disadvantageous effects do not show up until the system is implemented when fast, costly remedial action has to be taken. On this point McKinsey Associates (1968) state that in their experience, 'the issue of operational feasibility (will the system be used?) is far too often neglected until the new application is tried out in practice and perhaps found wanting – and this is the costliest kind of feasibility test'. They also criticise tests of economic feasibility:

> Since a company's computer resources are seldom equal to its computer opportunities, economic feasibility should almost always be a key criterion in weighing the merits of technically feasible projects. Yet it is frequently assessed rather casually, on the grounds that the important benefits are intangible, and intangible benefits can't really be evaluated.

A problem assessment of consequences, particularly in the human areas, is that the group most likely to be able to do this from an informed position is operational management, yet in our experience this group is often excluded altogether from early stages of the decision-making process or given only a minor sanctioning role. This neglect of human factors at the decision-making stage markedly increases the uncertainty of the implementation stage of a new EDP system.

The choice between alternative solutions

Up to now our examination of the processes of decision-making provides little support for early theories operating from a 'synoptic' model in which ends are given, all alternative courses of action are considered together with all their consequences, and an optimal solution is obtained on the basis of a simple preference function (Friedmann, 1967). The decision-making process certainly resembles this sequence of activities, but it is influenced by many variables not explicit in this model and search processes may be restricted by human cognitive limitations. Simon (1956), for example, points out that economists postulate much greater complexity in choice mechanisms, and a much greater capacity on the part of the organisation for obtaining information, than do psychologists; he believes that learning theories account better for observed decision-making behaviour than do the theories of rational behaviour. Existing theories also take little account of the requirement for reconciling different needs and interests as part of the decision-making process. These needs and interests may be economic, technical or human, and a solution that is high in one of these areas may be low in another, thus presenting management with a difficult choice.

Efforts are now being made to assist the uncertainty of the choice process by providing mechanisms for evaluating alternatives. A group set up by the National Computing Centre has directed its attention for a number of years now to developing improved methods for measuring economic factors. Both the authors of this book have written articles directed at drawing management's attention to some of the specific human factors to be taken into account when choosing a technical solution. Two choice areas attracting much attention at present are the technical and the political. It is not difficult to obtain information on technical alternatives and the choice between them may be easy. Similarly, political issues are usually given considerable weight, though these may not be brought out into the open. Solutions are rarely chosen if they are likely to offend any major power group within the firm. Leavitt (1958) points out that the final choice of solution is always mediated by the logic of the situation, and by the alternatives available and their costs, but also by issues of power and ideas of justice. Power politics often seem to be the dominant factor in innovative decisions and the one most likely to distort the economist's model of rational planning.

An important question for the researcher interested in the choice process is why a particular level of innovation was finally selected. Were the prime considerations here technical, financial, political or human – in the sense of learning limitations or resistance to change? We try to answer this question by examining the choices made by our case study firms.

The participants in decision-making

Decision-making is an exercise in coping with uncertainty but the participants in the process may themselves add to this uncertainty. Various factors may influence the decision-making behaviour of individuals. These include personal characteristics – the individual's desire for influence, status or even security, and the effect these have on his perceptions of the decision situation and on his choice of a solution. The decision situation can also exert an influence on the individual and the way he confronts uncertainty. For example, whether the decision is seen as being taken at a time of crisis or in a period of stability and development, either for the organisation as a whole or for an affected subunit? Again, influence from outside the organisation, such as a perceived demand for new products or the opportunities for greater efficiency and/or power offered by a new technology, may affect the behaviour of certain individuals in the decision-making group. In order to understand why particular decisions have been taken we need to understand the relative strength of influences of this kind and their impact on the searches for alternatives and consequences and the degree of novelty of the selected solution. When a decision is made by a group rather than an individual, and this is the case with most innovative decisions in the computer area, the degree of influence exerted by different factors becomes extremely difficult to identify. Cyert and March (1963) maintain that 'any alternative that satisfies the constraints and secures suitably powerful support within the organisation is likely to be adopted'. But what we need to know here is who exerts influence on a decision, how support for a particular solution is generated and which are the principal power groups involved.

Early studies of decision-making made few distinctions between decisions made by an individual and decisions made by a group. Even where group decision-making is examined the group is often regarded as a collection of people with a common interest who will all direct their energies to solving the problem. Conflicts of interest and the differential distribution of power and influence within a group are not usually examined in detail and may not even be recognised.

Arrow (1951) identifies this kind of problem when he says,

> A serious difficulty arises for theories of utility maximisation whenever a decision is to be made not by an individual, but by a group of persons. If the decision is to maximise utility not for the individual but for the group, then the utilities of the individual members must somehow be combined to provide a group utility (or expected utility) for each alternative. If the utility functions for one member of the group were identical with each of the other members of the group, then, of course, this problem would be trivial. But if, as is highly probable, utility functions do vary from one member to another, then the question of how utilities are combined in arriving at a group decision is an unsolved problem.

We have already accepted that a satisficing approach to decision-making is much more likely to take place than an optimising approach; another factor leading to a 'satisfactory' solution to a problem rather than to something more ambitious is the existence of a decision-making group which contains members with different interests. Taylor (1965) points out that

> group decision presents much less of a problem for the concept of satisficing. It will be expected that different members of a group approaching a choice will do so with minimum expectations with respect to a given value and may even see different values as relevant to the decisions to be made. If the rule in the group is that the decision must be *unanimous*, then satisficing would predict that the group would continue to search for an alternative until one was found that met the minimum standards of all members of the group with respect to all of the values that any member sees as relevant. If, however, the decision is by *majority* vote, the concept of satisficing would predict that the group would continue to search for an alternative until one was found which met the minimum standards of all members of the group for all values seen by some member of that majority as relevant. It should be recalled that aspirations of individual members may rise or fall as a function of experience in search. Values regarded as relevant may also change as a function of the experience in the group.

Top management

In any innovative decision concerning the use of computers the three participating groups are generally top management, the computer specialists and the management and employees of the user department. (In some European countries, notably Scandinavia, trade unions would also insist on being included in the group making decisions.) The role played by top management appears to vary considerably from firm to firm. Sometimes it sees computer developments as a part of strategic planning and takes a major role in the formulation of long-term computer policies. At the other extreme it may leave a great many of the decisions on how computers shall be used to its computer specialists and merely sanction these decisions and provide the necessary finance. Many criticisms have been directed at top management by observers such as McKinsey Associates on the grounds that it does not play a sufficiently major part in innovative decisions concerning the use of computers. Yet, as Barnard (1938) points out, a top management decision system is constantly expanding and in order to cover all relevant decision areas it may be compelled to leave a great deal of decision activity to its specialists groups. If this is the case then in order to reduce uncertainty in its own environment it must have confidence in the sagacity of its specialists groups, it must have some methods for checking on what they are doing, and it must have some means for controlling these activities.

In the case study described by Cyert, Simon and Trow (1956) it was interesting to observe how top management brought in outside consultants at intervals throughout the decision process in order to check on the solutions which its own specialists were putting forward. Our experience of the role of boards of directors in decisions for new computer applications is that many top managements do play a major part in the development of computer policy and in ensuring that this is closely related to other aspects of corporate strategy. They are likely, however, to rely a great deal on their specialists groups to bring problems to their attention and to offer ideas and recommendations though they will usually make the final choice of systems and hardware themselves. A study by Blankenship and Miles (1968) on management decision-making and its relation to size of firm, hierarchical level and span of control also found this to be the case. We found that a problem for the top management of some firms was ensuring that their specialists communicated to them unbiased information. There was a tendency for specialist groups to try to sell their top management the solution which they personally favoured.

Specialist groups

A characteristic of modern industry is the proliferation of groups of specialists within large firms. A major responsibility of these groups is using their specialist knowledge to assist the firm to take its more complex, non-programmed kinds of decisions. This means that many specialists have a particular responsibility for coping with the uncertainty of innovation. Davies and McCarthy (1967) define the objective of the innovator as, 'the joint study of society and of technology with the resultant identification of opportunity, its reduction to practice, and its adaptation into forms suitable for the commerce of his organisation'. Knight (1967) defines the innovator's problem-solving processes as follows: 'The innovator has a practical problem to be solved and searches for an idea or discovery that will solve it, or he has an idea or discovery and he tries to find a practical problem that his knowledge will solve.'

We have found that management services departments differ in which of Knight's approaches they adopt. Some management services departments wait until their line management comes to them with problems which they then help to solve – in other words they see themselves in what might be called a 'service' role. Other management services departments are more actively engaged in seeking out problems, see themselves in a 'ginger' role and believe that it is one of their functions to develop more innovative attitudes and interests in the management of their company; such groups are often found in conservative organisations unused to continuous innovation. Sometimes there appears to be a missionary zeal about their activities and they may talk in emotive terms of 'dragging our company into the twentieth century' or of other similar intentions.

Nevertheless few firms will be able to innovate without knowledge and Knight, in a study of the computing industry made between 1944 and 1952, found that firms which innovated had people with both a detailed understanding of the problem to be solved and a knowledge of the specific technology that led to each particular development. K. E. Knight (1967) also recognises that the innovator is not always guided by completely rational and objective considerations. He says,

> Emotional and social factors are likely to play a very big part in the behaviour of the innovator. Unlike creativity, innovation almost always involves social interaction. We should expect, therefore, that a person's beliefs about himself and his relationships with other people will be very important in determining whether or not a particular individual is an innovator.

Later on in this study we shall show how these emotional and social factors increased the uncertainty of one major innovative decision.

User departments

McKinsey Associates (1968) state that an important part of the decision process concerning new computer applications is testing their 'operational' feasibility. That is, will they be used by, and acceptable to, the management and staff who receive them? Very often this acceptability is related to the extent to which user departments are able to participate in decision processes concerning the proposed installation. We have found that the role played by user departments can vary dramatically. At the one extreme there is the department which controls its environmental uncertainty by identifying its own problem and asking its computer specialists for assistance in solving this problem; it then evaluates specialist department proposals in terms of its own financial, efficiency and human relations criteria; it participates in the design of new systems and takes responsibility for implementing these systems, once designed. In contrast there is the user department which is a passive recipient of innovation, playing little part in the decision processes and often implementing reluctantly and with serious doubts and anxieties.

When the user manager does play a major part in the decision process he has the same dilemma as the specialist, of attempting to take rational decisions while at the same time being strongly affected by his personal interests and emotions. Barnard (1938) suggests that a kind of dual personality is required in an individual contributing to organisational action – the private personality and the organisational personality. The employment contract made between an employee and an employer contains an implicit assumption concerning the willingness of the employee to abandon personal goals if these come into conflict with major organisational interests. In reality few people are prepared to do this altogether and a feature of organisational decision-making can be the attempts by key participants to reduce the uncertainty of their personal job situations by securing problem solutions which allow for a reconciliation of personal and organisational goals.

Many writers appear to assume that decision-making is solely a management function and does not occur elsewhere in the firm. Melman (1958) has been one of the principal writers drawing attention to the fact that the ultimate stage of any decision process resides at the bottom of a hierarchy because it is this group which decides whether or not to implement the decision, and the manner of its implementation. Although the staff of a user department may not be able to refuse to implement a new system of work they can decide that they do not like or want the system and that they will operate it in such a way that it cannot work effectively. They therefore have a key role in the decision-making process and a management that goes ahead with innovation without recognising this fact is embarking on a very high risk course. Even at management level the personal interests of the manager may influence the way he decides to cooperate with innovation and the extent to which he will direct his personal efforts towards organisational goals. The fact that innovative *action* – that is, the implementation and operation of new systems – is located at the bottom and not at the top of the decision tree would seem to be a powerful argument for the involvement of every level of staff in the decision process, even if the extent of this involvement is only ensuring that there is agreement with the changes that are to be made.

A major problem for the student of decision-making is distinguishing facts from values. Decision-making processes cannot be understood unless there is good knowledge of the personal values of key figures; unless also there is an awareness of group values and interests and an understanding of organisational values. Bross (1953) suggests that:

> The question of whether a given decision was a good or correct decision can never be answered on a purely factual basis, but only in terms of values. It is possible to consider whether the alternatives considered were those factually available, whether the consequences anticipated were those which in fact would have ensued, and whether the choice made was the one to be preferred given the individual's value system. If all these conditions were met then the decision can be said to be correct and good.

Issues of power and conflict as generators of uncertainty

The fact that in any innovative decision concerning computers a number of groups with different interests collaborate, means that this kind of decision-making cannot be systematically examined without consideration of issues of power and conflict. March and Simon (1966) have generally seen conflict as dysfunctional for rational problem solving because it introduces uncertainty into the decision-making situation. They define conflict as, 'a breakdown in the standard mechanisms of decision making so that an individual or group experiences difficulty in selecting an action alternative', and suggest that where conflict is perceived, motivation to reduce conflict is generated: 'We assume that internal conflict is not a stable condition for an organisation and that effort is consciously directed toward resolving both individual and intergroup conflict.' This implies that the participants in the decision process are able to recognise the existence of conflict and are prepared to bring it out into the open and to resolve it.

This would seem not always to be the case. Where there is an imbalance of power between participating groups conflict may be avoided or ignored in the hope that it will disappear; or there may be a pretence that it does not exist. An unwillingness to resolve conflict may arise because one of the groups participating in the decision process may actively seek to improve its own power position through the way it influences the decision that is being taken. Where this situation arises conflict is inevitable and may be irresolvable. The situation can only be eased by a higher group such as top management stepping in and preventing such a distortion of goals. Where there is willingness to resolve conflict this is likely to take the form of a reconciliation of different interests, with each group being willing to abandon some of its particular goals in the interests of achieving a decision acceptable to all.

Pettigrew (1973) has argued strongly that a more useful explanation of decision-making behaviour than that of satisficing is to look at the choice amongst alternatives as 'the product of the strategic mobilisation of power resources'. He sees the kind of decision-making in which groups with different interests and different levels of knowledge are involved as inevitably influenced by struggles for power, with each group striving to increase and/or maintain its organisational influence in order to secure a more strategic position in the organisation.

Cyert, Dill and March (1967) after analysing the study described earlier concluded that 'expectations were influenced by hopes, wishes and the internal bargaining needs of subunits in the organisation'. Because information on the consequences of certain courses of action was hard to obtain there was a conscious manipulation of expectations by different groups. We too have found many disappointed expectations once new computer systems are implemented. These expectations have been acquired during the decision processes as specialists sought to influence user departments to accept a particular kind of solution.

Stress and skill in decision-making

The factors mentioned in the last section can make innovative decision-making particularly stressful for the participants. In addition to the high financial risk associated with the introduction of novel solutions there is the inevitable conflict of interests and the certainty, if the balance of power between participating groups is more or less equal, that a compromise solution will have to be accepted. This is usually particularly painful for the technical specialist involved, for his technological utopia is rarely found acceptable by other groups. Barnard (1938) has pointed out that making decisions can be unpleasant; there is exhilaration if decisions turn out to be correct, depression if they are wrong, and frustration if there is uncertainty about their success. Major innovative decisions are generally expensive and difficult to change and participants in the decision process are likely to have to live with their mistakes for a very long time.

Barnard suggests that men try to avoid making decisions and that their capacity to do so is narrow although it can be developed by training and experience. The skill of decision-making lies not only in the selection of a course of action from a number of alternatives but also in ensuring that this course of action is implemented. For a manager this is likely to involve seeing that people lower down the company hierarchy make effective decisions, and this may mean the design of efficient decision-making systems. Yet, as Simon (1965) points out, this is a different skill area and there is no guarantee that the man who is a good personal decision maker is equally good at creating communication systems which enable others to make good decisions.

Dufty and Taylor (1962) have identified five types of communication associated with decision-making, similar to those previously described by March and Simon. First there is communication associated with non-programmed activities; for example, discussion with a trade union. Secondly, there is communication to establish search and implementation programmes, and this may involve conferences between different groups of people. Third, there is communication to provide the necessary data for the execution of programmes. Fourth, there is communication to initiate programmes; for example, the communication of decisions, the issuing of instructions. Fifth, there is communication to provide information on the results of activities – how things have worked out. These different kinds of communication are an integral part of the decision process and we look at them further when we examine our case study firms.

The decision environment

In the last section we examined the uncertainty brought to the decision-making process by the various participants and their different personal and group interests. The external environment of the firm may also increase or reduce decision-making uncertainty. The student of decision-making cannot fully understand the rationale behind organisational decisions unless he is fully informed on the nature of the interaction between an organisation and its environment. Similarly, managers will not make decisions which enable the organisation to respond better to pressures from its market environment unless they are fully aware of the nature of these pressures. Friedmann (1967) tells us that 'the *best* proposal will always be the one which in the accomplishment of ends, and in the setting of them, takes full cognizance of the social context'.

A problem for all organisational decision makers is that the environments in which organisations exist are today changing at an extremely rapid rate and becoming very complex. Managers are subject to continuous change in product and labour markets and to the frequent opening up of new technical opportunities offering improvements in the way they make their products or run their businesses. Thus the modern organisation is presented with the problem of behaving more or less rationally, or adaptively, within rapidly altering product, labour and technical environments (Simon, 1956). In order to comprehend decision-making processes an understanding of environmental influences and the nature of the exchanges between an organisation and its environment is therefore essential.

Emery and Trist (1965) have set out the different types of environment in which organisations may exist. The simplest kind of environment is the *placid* one in which goals are few and relatively unchanging and the organisation has only to adopt the simple tactic of learning how to respond successfully to its environment in order to achieve and to continue achieving its goals. At the next level there is a more complicated but still relatively placid environment in which goals are interdependent and the organisation has to develop a strategic response to its environment if it is to achieve these linked groups of goals. Thirdly, there is what Emery and Trist call the *disturbed–reactive* environment in which there is more than one organisation of the same kind and these organisations are in competition with each other. In order to achieve its goals the organisation has therefore to respond to the behaviour of these other organisations. The most complex environment Emery and Trist call a *turbulent field*. This is a highly dynamic environment and the dynamism is a product of the interaction of different variables within the environmental field. For example, economic factors are today constantly interacting with technical, human and political factors, thus presenting an organisation with an endlessly shifting range of pressures.

Most organisations now embarking on major technological change are operating in this turbulent field type of environment and this requires them to develop decision-making processes which are competent to respond to changing environmental pressures. Here different authorities offer different solutions. Emery and Trist (1965) suggest that one way of meeting environmental uncertainty is by the development of 'values that have overriding significance for all members of the field'. That is, when faced with problems the groups concerned are broadly in agreement on how they should be solved. Stafford Beer (1972) has a different kind of solution. He believes that the modern organisation needs to set up groups whose major function is the constant scanning of the turbulent environment in order to identify and bring to management's attention things to which it should be responding. Beer believes that all decision-making must be associated with excellent monitoring and feedback mechanisms so that changes in the environment can be instantly identified and decisions modified to take account of them. He suggests that any decision regarding the solution of a problem taken at one moment in time and implemented unaltered at a later date will surely be a wrong decision because it lacks a response to changes in the decision environment.

A major problem for management is how it can best organise itself when making decisions so as to be promptly aware of when and how significant factors in the environment are altering and giving rise to a need for adjustment. This relates back to the problem to which we referred earlier of when to stop the search for alternatives. It might be argued that a too sensitive response to environmental change will result in no decision ever being taken because the situation will not stand still long enough for a completely acceptable decision to be implemented.

Conclusions

The points we have made up to now demonstrate the uncertainty and complexity of the decision-making process and show the problems involved in making major innovative decisions. It might be argued that innovative decision-making is so onerous and so risky that few people would be prepared to be associated with it. Fortunately for progress this does not appear to be the case. A saving feature for the individual or group responsible for innovative decision-making is that it is often difficult if not impossible to identify when a wrong decision has been made. Substandard decisions may never show up because of the impossibility of knowing what would have happened if a different kind of solution had been implemented. Also, the timespan between a decision being made and its subsequent implementation may be very long. Stafford Beer (1966, 1972) would argue that providing a decision and its implementation move the organisation generally in the direction that it wants to go, then the decision has been a reasonable one and should not be criticised. Others would suggest that the art of good decision-making is negative rather than positive. It lies in *not* making decisions which are not pertinent at a particular moment in time, in *not* making decisions prematurely; in *not* making decisions that cannot be made effective, and in not making decisions that others should make (see Barnard, 1938).

It is clear that the manner in which a decision is made greatly influences the decision reached, and that where a number of different groups are associated with the decision-making processes conflicting interests may have to be reconciled through some form of bargaining (Friedmann, 1967). In chapter 5 we argue that rationality in decision-making is frequently displaced and corrupted by internal politics, these politics being a product of sectional and individual interests.

PART THREE

Uncertainty and the innovator

4

Uncertainty and innovative decisions – four case studies

Computer based solutions to organisational problems are frequently a response to uncertainty in the firm's internal or external environment. There may be a recognition that the firm is slipping behind previous standards of administrative performance, or a belief that the creation and implementation of new production or marketing goals can provide opportunities for improvement in areas such as customer service and product or market performance. Nevertheless responding to uncertainty by accepting a computer based solution to a business problem will itself generate new and stressful kinds of uncertainty as the firm strives to assimilate and use this form of technology. There may be uncertainty about which problems to tackle first with the computer – what will yield the surest and fastest reward? There will be uncertainty as groups affected by the innovative process perceive their status and power to be threatened by the proposed change, and there will be uncertainty as individuals attempt to find out the implications of the new technology for their work, prospects and job security. In this chapter we examine the kinds of uncertainty experienced by a number of different firms when introducing large-scale computer systems, and their methods for coping with this uncertainty.

In all these firms the introduction of computers was much more a response to a desire to realise new goals than to feelings of dissatisfaction with current performance. A technology had arrived which made possible a level of operating efficiency that could not be achieved with existing or improved manual systems.

The approach to innovation took a broadly similar form in each firm although the methods for handling the uncertainty generated by change were very different. The stimulus to use or make more use of computers was one of two kinds. Either internal or external pressure for operational change led the firm to consider a computer based solution to its problems, and to make a decision on which problem or problems relevant to this change it should tackle first. Or, if the firm already had considerable computer hardware and wished to utilise this to a greater extent, then problems must be identified which would be effectively solved using a computer. In the first approach the problem took priority and a consideration of means for its solution would be likely to include non-computer as well as computer solutions. In the second approach existing computer hardware was the major influence and problems had to be sought out in order that this expensive technical investment could be more fully utilised.

Whichever of these routes to action was used, the first step in computer innovation was always the specification of problems for solution and we were interested in finding out how our firms handled this. Did they, for example, leave problem selection to senior managers at board level, who had the insight and imagination to perceive how computers could best be used. Did they allocate responsibility for identifying problems to their line managers? Or did they create new groups of specialists, part of whose occupational role responsibility was to identify problems and rank them in terms of urgency?

Once a major problem-solving goal had been formulated, the next step was to specify the selected problems clearly and in detail, and this required the setting of subgoals without which the overall system goal could not be achieved. We wished to observe how this subgoal setting was carried out in our case study firms, how successful it was and the extent to which goals set at an early stage of the innovation process were altered over time as planning proceeded.

The act of setting a strategic goal or goals and delineating this through the specification of subgoals was a procedure likely to generate considerable uncertainty. We wanted to know what form this uncertainty took. Did it appear in a clash of interests and values between important individuals in the firm? Certain members of the board might, for example, believe their power position to be threatened by a particular use of computers. Was there also uncertainty among and between different occupational groups – especially computer specialists and people located in the intended user department? Did the uncertainty take a technical rather than a human form because of a lack of knowledge of how to choose appropriate hardware for solving the selected problem? Was there uncertainty because computer based innovation was new to the firm and it lacked experience of how to order and evaluate available information?

Hypothesising that when major uncertainty exists in an organisation, steps are taken to reduce it, we wished to see how our firms handled the uncertainty generated by their decision to solve problems using a computer. How did they 'manage' this complex change that they were introducing? Recognising also that uncertainty is likely to lead to conflict we were interested in how the firms resolved and contained this conflict. What mechanisms did they use to mediate between different interests and achieve a reconciliation? Were they aware of, and did they cater for, the workings of the firm's internal political system during a period of major change? These are all issues integral to an understanding of the dynamics of innovation.

The four users

The four large-scale computer applications which we studied in the hope of obtaining answers to the questions set out above were located in different kinds of organisation; two in manufacturing industry, one in distribution and one in government service. These computer applications were a production control system, a sales distribution system, an order processing system and a system for dealing with wage payments and personnel record data, which would eventually provide information for manpower planning. Each of the host organisations had a long history of computer usage, and all were experienced in dealing with this kind of change, but each of the applications was the first computer system in the chosen user area. We shall begin by showing how each of the applications originated and the computer history which preceded it.

Government service: AEK

The government service organisation, which we shall call AEK, was introducing a pay and personnel records system. It was a large unit but was part of a much larger government organisation which began to use computers at a very early stage in the development of this technology. If we examine AEK's clerical procedures in the 1950s, we find the payment of wages and salaries decentralised to a large number of local offices, with a tremendous amount of work for the local clerks who ran the pay system. Traditionally pay was calculated by a double ledger, 'fair and rough' method with two clerks, a senior and a junior, working each pay account separately and then comparing results. This method required the use of a large staff of pay clerks and it was becoming difficult to maintain because of recruitment problems. By the end of the 1950s considerable thought was being given to the possibilities of a new approach and two courses of action were being considered. These were (a) whether to centralise the system; (b) whether to computerise the system. A number of senior managers were very much in favour of using computers and their enthusiasm had an influence on other staff. The main arguments both for reforming the system and for computerisation were:

1 problems in recruiting clerks and thus difficulties in maintaining the adequacy and efficiency of the manual pay system;
2 pressures on the manpower costing and forecasting sections to produce more accurate information more quickly;
3 keeping in line with other government organisations who were more advanced in their use of computers for pay and personnel tasks.

In the early 1960s it was decided that a study should be undertaken to examine the feasibility of using computers for the compilation of pay and for other personnel activities such as personnel records and manpower planning.

Manufacturing firm: Falcon Ltd.

The first of our two manufacturing firms, Falcon Ltd, moved into electronic data-processing in the late 1950s, buying its first computer in 1960. This computer was used for compiling production records and proved very successful. Much of the computer output at month and quarter end was fed to the Accountancy Department, thus paving the way for the adoption of computerisation. At this time the clerical workload in the company was growing rapidly and the firm was using Hollerith machines for sales statistics, costing and bonus calculations. Falcon was part of a larger group and in the early 1960s the local Board suggested to the Group Computer Panel that a commercial computer should be acquired; this was agreed and two computer manufacturers were asked to make proposals for transferring the existing Hollerith work to a computer. The tender of one of these manufacturers was accepted and this set the pattern for future hardware purchases as the firm continued to buy from this manufacturer thus avoiding the decision problems associated with the choice of hardware that were experienced by one of our other firms.

The department most intimately concerned with the acquisition of this computer was the General Office. This department controlled the firm's office services, machine maintenance and document reproduction, and also its Organisation and Methods section and the Hollerith machine centre. By 1964 the firm was underway with a programme of rapid computerisation. Wages and salaries were soon to be transferred to the computer and plans were being made to transfer production scheduling. The General Office's order invoice systems were also to be transferred and the application we were to study in detail (distribution services) was planned during this time. By 1968 all these computer projects were either implemented or their formulation was well advanced.

Manufacturing firm: Grant and Co.

Our second manufacturing firm, Grant and Co., was also part of a larger group and the Group Computer Centre assumed responsibility for planning and implementing the firm's first computer application, a production control system. The implementation of this system took place in the late 1960s. As our research was concentrated in Grant and Co., while early decisions on the choice of hardware and the broad systems approach were taken at Group Head Office, we do not have a great deal of information about the first stages of this project. Our period of observation began when the Head Office systems analysts arrived at Grant to carry out the feasibility study.

Distribution firm: Poultons

Our distribution firm, Poultons, had been a pioneer in the use of computers in its industry. Early applications had been principally in the field of stock control but over a number of years the Board of Directors and the head of Management Services had given a great deal of thought to identifying further areas of the business which could usefully be assisted by computers. They had come to the conclusion that the efficiency of the firm's central office could be improved if order-processing and accountancy procedures were computerised and it was this application which we were able to study in detail.

The firm had problems with this office because of a shortage of female clerical labour in the area and senior management believed that the transfer of a major section of the work to a computer system would greatly reduce the required number of clerical staff. The Board later decided that it would extend its computer operations to cover all Central Office accounting procedures. The decision on the kind of system to be implemented, the choice of hardware and the design of the system took five years; the system was finally implemented in March 1972.

All these organisations initially approached computer usage from the angle of a recognised business problem, although in both Falcon and Poultons the fact that computer hardware and systems design experience already existed facilitated and accelerated the solving of other problems by means of computers. The early decisions to innovate provided the technical means and knowledge for further innovation and led these firms into long-term computer development programmes. Because each firm now had the technical means to solve many of its business problems it was led into an investigation of problem priorities – which of the firms' activities would acquire the most benefits from computers? Which computer applications would most logically follow from existing ones? This was an interactive process with the recognition of a problem area usually preceding a decision to attempt a computer based solution. On occasion, however, it was perceived that valuable and expensive computer hardware was not being fully utilised and a search for additional problems took place. We did find that a major stimulus to computer innovation was the presence of imaginative, entrepreneurial managers who were prepared to take risks. In Falcon Ltd these men were located both on the Board and in the newly formed computer department. In Poultons, a forward looking head of Management Services acted as a great stimulus to risk taking innovation. The use of computers in Falcon and AEK was also encouraged by intelligent line managers who recognised the potential of computers for improving the efficiency of their departments and pushed hard for this kind of technical assistance to be provided. In Grants and Poultons we found the opposite situation with the stimulus for computer usage coming from the Board and the Computer Department and not from line management. Here line managers were nervous of computer systems, seeing these machines as leading to a great deal of uncertainty with which they might be unable to cope.

Reduction of uncertainty through a favourable environment

One way of reducing future uncertainty is through the setting up of a well structured, viable decision-making environment. Managers, when considering a new or extended use of computers are faced with answering such questions as, 'Here is a problem, shall we continue to live with it or try to solve it?' 'Should we solve it now or later?' 'Can we specify the problem in such a way that it can be solved?' 'How much is it worth spending on the solution?' All these questions were asked by our firms, though some of them were not easy to answer. For example, it is common practice to make a choice among a number of problem areas by evaluating financially the payoff which will result from tackling one problem as compared with another. In the computer area this kind of economic evaluation is still very difficult. The approach adopted by our firms was for problem selection to be a board or senior management decision, with advice on how the selected problem could be solved coming from the computer specialists. For reasons which we discuss in chapter 5 this advice was not always disinterested and objective but could reflect the personal interests of the technical experts. We found that senior management in all our firms had checking procedures to vet the opinions of their experts, before they took a decision involving considerable financial outlay. These checks took the form of a careful scrutiny of Computer Department proposals at Board level, the employment of outside consultants to check internal experts and, in at least two of the firms, the formation of special committees with members drawn from top management and user areas.

The government service organisation, AEK, had an extremely comprehensive committee structure, with new committees being set up whenever a difficult problem area or a conflicting set of interests was identified. Broadly speaking, these committees carried out the functions and activities which Shafer (1967) recommends as important for all central planning. These, removed from a national planning context, are as follows:

1 The representation of major interest groups
2 Technical appraisal of current trends and formulation of aggregate goals
3 Preparation, coordination, or review of major projects
4 Mediation among competing groups and leaders
5 Overall leadership
6 Central financial management
7 Handling critical issues of the moment
8 Symbolic or ceremonial functions

Such activities were not carried out by all committees but the responsibilities implied in them were accepted, and allocated to different groups.

AEK had had an Organisation and Methods department since 1942 and a strong interest in computers was formalised in 1956 through the setting up of a high level computer policy-making committee. This committee's brief was to initiate appropriate studies and monitor their progress and it included specialist advisers from sections such as O and M, together with the heads of potential user sections – Accounts, Supplies, Manpower and Manpower Statistics. A working party on accounting was already in existence and for a number of years this had been studying ways of reducing the workload of the clerk concerned with the handling of pay.

In the early 1960s a decision was taken to make a detailed study of the possibility of transferring pay accounting, manpower statistics and personnel records to a computer. A four-man study team was created with members from Accounts, Statistics and O and M sections – the user departments which would be affected by a computer system. Their brief was to consider:

1 the feasibility of a computer for pay accounting, including a comparative appraisal of alternative approaches providing for different degrees of centralisation;
2 the feasibility of a computer scheme for other personnel work including the maintenance of personnel records and the provision of data for manpower statistics.

A steering group was set up to monitor progress and it was decided that the study team should concentrate initially on the pay area. The study team had an information-gathering rather than a goal-setting role and its immediate aims were:

1 to discover any factors peculiar to the Department (as opposed to other government departments) which might preclude the introduction of EDP;
2 to assess the appropriate degrees of centralisation;
3 to devise in outline a suitable computer system covering the pay aspect (in so far as this was practicable in isolation from other tasks which might later be included in the overall scheme);
4 to assess the economics of such a system (subject to the same proviso as in 3).

The team began this information search process by familiarising itself with pay procedures. The team members examined the pay systems of other similar departments and they attended a number of EDP courses. As the search and decision processes progressed other study groups and committees were set up to investigate specialist aspects of the proposed computer system and to evaluate the proposals of the study team. In addition a number of existing high-level policy groups had a monitoring and vetting role.

This approach to planning and decision-making resembles the model of rational decision-making set out in chapter 2, although it is even more comprehensive, incorporating as it does mechanisms for searching, checking, the reconciliation of different interests and the division of responsibility. The planning group had a well defined specific objective, they had groups of experts available to search out alternative ways of tackling the problem, and they had a clearcut search procedure. In addition the number of formal groups associated with the decision process meant that decisions were only taken after good evidence had been submitted and accepted. The advantage of this approach was that, because all interested parties were involved in decision-taking, the decision process appeared to be less affected by major uncertainties than was the case in our other firms. As in Falcon Ltd, an early decision on the make of computer to be used reduced the length of the hardware search. In addition, the fact that all groups involved in the decision processes had clearly defined and well understood responsibilities seemed to lessen the chance of internal politicking and attempts to influence the decision process to suit particular vested interests. The early and direct involvement of top management was of great assistance in ensuring that major changes in policy did not occur before the final decision was taken. It might be argued against this approach that so many formal search, evaluation and decision-making bodies will slow the planning process down and prove expensive in time and money. However we have no evidence on this point and it is impossible to assess if a simpler decision network would have led to faster results.

Falcon Ltd, the first of our two manufacturing firms, adopted a similar approach as a means for controlling and reducing uncertainty. In the early 1960s a computer group was established and attached to the General Office with a brief to prepare programmes for a new computer that would take over work then done on a Hollerith machine. This team was recruited internally and its members were drawn principally from the Accounts and General Office Departments. None of the team members had any previous experience of computers and so a training programme had to be arranged.

Initially this group was under the joint control of one manager from each of the departments which had contributed to the composition of the team, the Accountancy and General Office Departments. It reported to a steering committee consisting of the Deputy Chief Accountant and the Head of the General Office, who in turn were responsible for the whole computer project to the Finance Director. The arrangement, although logical, caused the Head of General Office to have a split responsibility as his department was responsible to the Personnel Director for all its other activities. In 1962 Computer Services had the following organisational position:

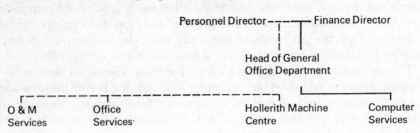

O and M Services and the Hollerith Machine Centre came under the control of the Personnel Director while Computer Services were under the Finance

Director. This split in control was recognised as illogical and in 1964 O and M services became the responsibility of the Finance Director. By this time the Hollerith Machine Centre had gone and its work had been taken over by the Computer Department. Data used by the disbanded Hollerith Centre were handed over to the Computer Department and the Hollerith Machine girls became console and punched card operators. The computer group had now become a fully fledged computer services organisation.

The Finance Director next recruited a long-term development group to explore possibilities for further computerisation throughout the company and he ordered a second computer. Now that the Hollerith work had been successfully transferred a number of programmers were asked to develop a computerised wage payment system. An advantage of tackling wages next was the fact that Accountancy, of which the wages section was a part, and the Computer Department both came under the same director. There were therefore few problems in deciding what the computer group should or should not do since the same man was ultimately responsible for both functions. The necessity for reconciling the possibly conflicting interests of two directors was avoided and this area of uncertainty eliminated. Thus although no standard procedures for dealing with computerisation had yet been developed, a viable organisation structure had already emerged.

The design and implementation of the computerised wage payment system and the creation of the long-term development group marked the end of the first phase of computer developments at Falcon. All the work which had previously been done on the Hollerith punched card system had been transferred to the computer and all wages were being calculated by the computer. A feasibility study for computerising work in the Distribution Services Department had been completed and one section of this department went onto the computer in the early part of 1965. Meanwhile the long-term development group had been recruited and trained and they began their work with the remit to devise and introduce an integrated information system, a task expected to take some time.

The Computer Department was now well established and well integrated into the company. It became increasingly in a position to choose which business area to enter next (of the several who invited it in) whilst the individual analysts were finding it easier to sell intangible benefits to managers already convinced of the advantages of computers. These managers perceived EDP systems not only as manpower savers but also as important aids to business organisation and management decision-making.

An analyst said of the department of one of these managers, 'because they were as enthusiastic as they turned out to be we were able to concentrate less on clerical savings than would have been the case if we were proving that a computer system was economic in a department that did not want it. In other words, they would accept intangible benefits as part of the justification.'

These two firms provide examples of well planned attempts to reduce uncertainty through carefully thought out decision media in which problems are selected and computer goals set through joint decision-making by top management, system users and computer specialists. In addition to the technical experts located in established computer departments, each organisation set up a series of policy making groups and working parties to examine and control new computer projects.

In AEK there were the computer policy-making committee, the working party on accounting, a number of study teams and a steering group: all worked closely with each other and acted as discussion forums and checking mechanisms for particular proposals. Because AEK was a government body, there was in addition the powerful financial check exercised by the Treasury requiring a strict economic evaluation of computer proposals. In Falcon Ltd there was a similar though less formalised committee structure. In addition to the permanent Computer Services Department, special study groups and steering committees were set up to examine specific projects. As in AEK there were external checking mechanisms, for computer projects in the commercial area could not be implemented without the influence of a Group Computer Panel, and financial expenditure had to be approved by Group Head Office.

Both AEK and Falcon Ltd involved user management in the decision processes at an early stage and so avoided causing user uncertainty through lack of information about computer policy. No decisions were taken without user agreement and this approach was perceived as contributing to the eventual success of computer projects. In Falcon Ltd, this early involvement made user management so interested in the possibilities of computers that ideas for new systems originated with them rather than with the Computer Department.

In both organisations computer innovation was stimulated by the existence of individuals who were enthusiastic about exploiting this new technology. In AEK these were a number of senior staff in user areas with responsibilities for activities such as manpower planning and personnel records who were dependent on accurate information for the performance of their duties. In Falcon Ltd the chief stimulus came from the Finance Director who not only had authority over the Computer Services Department but also those commercial areas where computers could be most usefully applied. His dual responsibilities guaranteed that both computer specialists and user management would be keenly interested in proposed computer developments.

Our two other firms approached decision-making in a different way. Committees and working parties were organised, but not until a much later stage in the decision process. Grant and Company, our second manufacturing firm, set up a number of working parties once the decision to introduce a computer had been taken at Group Head Office and the systems design team had arrived at the subsidiary firm. However, local managements' lack of knowledge of computer systems meant that their decision-making powers were gradually eroded and taken over by the computer specialists loaned by Group Head Office. In Poultons working parties were not set up until the computer system had been designed and was almost ready for implementation. Users were therefore only involved in decisions concerned with implementation

policies and played no part in the design of the system. This approach appeared to reduce uncertainty for the computer group, who then only had to
interact with top management, but to increase it greatly for the user department. In these two firms we found that a lack of a clearly defined and structured decision environment led to communication gaps between computer
specialists and top management and between computer specialists and user
departments. Computer specialists, perhaps because of their isolation, tended
to develop vested interests in particular kinds of solution.

Thompson (1967) has argued that in any complex organisation, power is
dispersed, and that for the organisation to be decisive and dynamic, the dispersed power must be reflected in and exercised through an inner circle. In
AEK and Falcon Ltd this inner circle took the form of a committee structure
organised so as to represent all interested parties. This reduced conflict by
facilitating consultation and communication. In Grants the inner circle
reverted from a coalition which included users to the computer group alone,
because a lack of knowledge made user management ineffective decision
makers. This led to an increase in uncertainty and to communication breakdown. In Poultons the computer group tried to keep power in its own hands
by providing top management with selective information and not allowing user
areas to participate in decisions until a very late stage in the planning process.
They were partially successful in doing this although the fact that they did not
have a goal-setting coalition with the Poulton Board led to unexpected changes
in goals by the directors which increased the computer group's uncertainty.
At the same time the non-participation of users in the early decision processes
generated a great deal of emotional uncertainty in the user area.

Thompson (1967) suggests that concentrated power may be found in the
technical core of those organisations which have standardised, repetitive
activities, and which have succeeded in isolating or buffering those activities
from environmental fluctuations. This does describe the Poulton situation for
it was a tightly structured firm operating in a stable market situation. Thompson also maintains that even in this situation power will be diluted if technical
complexity exceeds the comprehension of the individual (we would add here,
'or specialist group'), when resources required exceed the capacity of the
individual (or group) to acquire and when the organisation faces contingencies
on more fronts than the individual (or group) is able to keep under surveil-
lance. In Poultons the factor that most affected the ability of the computer
group to control decision-making was the role of the Board, for it was the
Board that made the financial decisions. However, through the kind of information which they fed up to the Board, the computer group were able to
influence the way this financial power was used. Two factors in Poultons
assisted the location of a great deal of power in the computer group. The first
was that they, and only they, possessed the necessary technical knowledge to
make technical decisions. The second was that user areas were passive and at no
time sought power by demanding that a coalition be formed to make decisions
and that they be included in it.

Goal-setting as part of the decision process

Cyert and March (1963) claim that coalitions are formed through a process of bargaining which determines both their composition and their general terms of reference. We would argue that once such a coalition *is* formed then the decision activities it participates in are also greatly influenced by bargaining amongst the different interests represented in the coalition. In this section we discuss the kind of goals set for the computer applications we were observing, and try to identify how they were set and the extent to which goal-setting represented a reconciliation of different interests.

A feature of all our firms was the dynamism of the goal-setting process. Goals were not set when planning began and rigidly adhered to until the computer system was finally implemented. On the contrary there was a continual feedback process between goals and information, with goals being redefined as new technical information was obtained or as factors altered in the firms' external environment. New developments in technology meant new system possibilities and a higher level of problem solution than had previously been envisaged and goals were altered to take advantage of these. Similarly, the board of the firm might receive information about computer developments in a competitor and this information would stimulate a change in goals. Goal changes were principally related to modifying the size of the computer system and the functions to be covered by it. New decisions were taken on questions such as 'How large a problem shall we attempt to solve?' This rethinking would involve a change in the position of system boundaries with, as a rule, more business functions being computerised. It could also involve the purchase of additional or different computer hardware which in turn would have an impact on system design. The process of goal setting was therefore much more interactive than is implied in the rational planning model (see Fig. 2.1 page 28). In this model goals appear to be fixed at an early stage and the role of the decision maker is that of selecting the best method for achieving these goals. The behaviour of our firms fitted more closely to Stafford Beer's (1969) description of effective planning. Beer argues that present planning philosophy is wrong because it assumes that a plan is formulated in advance of events, with means then being devised to effect it. The plan must next be vigorously prosecuted to execution with its success being measured in terms of the extent to which the plan is implemented as originally conceived. Beer believes that this is a recipe for disaster because the timelag between goal setting and goal implementation will mean that goals set at time A will have lost some or most of their relevance at time B. He believes that successful planning requires the organisational capability to abort plans on a continuing basis. In Beer's words 'adaptation is the crux of planning'.

In AEK we found that broad and indeterminate goals were set at the beginning of the planning and decision-making process, with goals being altered and made more specific on the basis of information provided by the groups responsible for the information search processes. For example, in 1959 the working party on accounting was given revised terms of reference. From looking at ledger-keeping only they were asked to review 'the current system of pay accounting in the light of the need to reduce the amount of work . . . in connection with pay accounting *as well as* the desirability of making maximum use of the most modern methods of mechanised accounting . . .' Between 1960 and 1962, as a result of investigations related to the new terms of reference, the working party recommended that the use of computers for pay work should be considered by a small feasibility team to establish if a full scale EDP study on the system would be worth while. It was also suggested that there should be an approach to the policy-making committee to ascertain if other personnel tasks should be included in the investigation. In 1962 this team produced an interim report suggesting a scheme which would replace ledgers with a computer produced pay sheet. The rough costing made suggested that such an approach would just break even and that in financial terms it was 'finely balanced'.

In 1963 the steering group formally suggested that the team should go ahead and investigate statistics and personnel records. The team members now began to consider the whole scheme in more detail and attempted to calculate the length of a conversion stage from a manual to a computerised system. They considered too the question of liaison and possible integration with another government department which was engaged on a similar scheme to computerise pay and other personnel tasks. The team made a second report on all these points to the steering group which forwarded it to the Computer Policy-making Committee, together with a paper which made the points that:

1 the study should still be considered as at the feasibility stage, as an economic case for the system had not been proven; and
2 before any decisions could be made further study would have to be made with regard to finding a speedier conversion time, deciding on a location for the computer, and also to examine the other government department computer developments and to consider the integration of both pay systems on one machine.

This question of sharing with another government department seems to have arisen through the information seeking activities of the team. The steering group thought that this was unlikely to prove advantageous, and in the event no sharing took place, but suggested that a small group be set up to look into the possibilities. The study group would continue to examine the conversion problems, and it was also suggested that a further panel should be set up to consider the possible location and manning of a computer centre.

It can be seen that goal-setting in AEK was a cautious, interactive process in which a set of viable goals was slowly and painstakingly built up as a result of extremely comprehensive search processes. At each decision point there was rigorous checking by groups other than those which had carried out the investigation and these groups had to approve the recommendations of the 'search' groups before any system goal was accepted. Goals were not formulated at the beginning of the planning process, instead questions were asked

which would lead the searchers in particular directions. Specific goal formation took place after many investigations had been made.

Falcon Ltd, took a similar approach. This firm followed the strategy
approved of by Beer (1969) who states 'the ostensible object of planning – a
realisable event – happens from time to time as a fall-out of the planning process which passes it by'. Falcon believed that planning for computers should be
long-term and closely related to other aspects of the firm's corporate strategy.
This meant that as corporate strategy was altered to meet new factors in the
firm's environment, so computer policy was adapted to fit the changed situation.

Fig. 4.1 Interactive search model

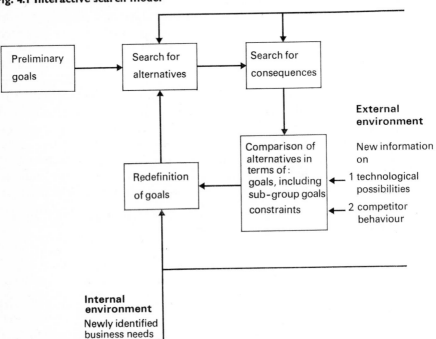

Falcon's investigations into the feasibility of electronic data-processing
began in 1957. From that time on the firm engaged in a very ambitious computerisation programme with the result that it had to face the complexity and
uncertainty of making decisions for a number of computer projects simultaneously. Once the work done by the Hollerith Centre – sales analysis, labour
and material costs and bonus calculations – was on the computer, the computer group began to consider other applications with a view to loading the
new computer to its full capacity. Capital costing, wages and product costing
suggested themselves and feasibility studies were made for these and also
for the work of the Distribution Department. Each application was, wherever
possible, built to fit with and extend an existing application so that system
incompatibilities were avoided.

Poultons followed the same approach, but to a more limited extent. Although the computer department constantly urged the Board to formulate a long-term computer strategy, projects tended to be looked at individually; on their merits for solving particular problems rather than as an integrated part of a long-term plan. In Grants the application which we studied was their first and a long-term computer plan did not exist.

The two models shown in Figs. 4.1 and 4.2 set out the approach of our firms. The first model shows how the search process itself leads to a redefinition and

Fig. 4.2 Technological stimulus search model

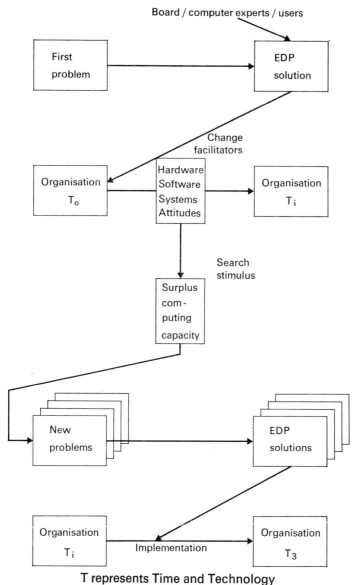

T represents Time and Technology

reformulation of goals. The second model, developed by Hedberg, shows how once a firm has acquired a computer there is a likelihood of it having surplus computer capacity and this, in turn, leads to a search for new problems which would be amenable to solution by computer.

The generation of uncertainty

We are arguing in this book that a problem for a management introducing major technical change is to control and contain the uncertainty generated by the change process. In this section we try to throw some light on the nature of this uncertainty and on its causes. The aspects of planning and decision-making which we have discussed in the previous sections – creating an environment for taking decisions and the formulation of system and planning goals – though used by management as a means for reducing uncertainty in the decision process may also generate uncertainties in other areas, for example, in the balance of power in the organisation and in interpersonal relationships. Exclusion from search, evaluative and decision-making groups will always generate uncertainty unless such groups set up excellent mechan isms for communicating their findings, permitting discussion of these and taking account of majority opinion. By virtue of their functions these groups are able to acquire information and knowledge, commodities which are in scarce supply during periods of change and which are, in consequence, valued very highly by those who do not have them.

Other factors found to generate uncertainty at the personal relationship level were clashes of individual and group interests and values on the nature of appropriate planning and systems goals and on the means to be used for implementing the new system. Clashes of interest were related to personal and group needs for power and security. Staff and management in user departments often viewed a proposed computer system as a threat. Managers might think that the introduction of a new technically based work system which they did not fully understand would erode their power and influence in the firm. Such a system would, they believed, make them less competent to run their departments and more dependent on the knowledge and expertise of the computer specialists. Managers and their staff sometimes feared that the new computer system would jeopardise their personal security in the firm, either because of the need for learning new ways of doing their work, which might prove difficult, or through the possibility of redundancy because of the labour-saving potential of the computer. Within the computer department, too, there could be clashes of interest as individuals and groups located there favoured different kinds of solution to the systems problem.

Clashes of values took a rather different form and were related to perceptions of how problems should be solved. The technical computer man tended to go for a high level technical solution which fully exploited the potential of the computer hardware. The user department, in contrast, frequently wanted more attention to be paid to human needs and preferred the computer system to be viewed as a tool for users, rather than users as adjuncts to the computer system. We found this question of values an important one. Innovative decisions could only be understood if the values of the participants in the decision process were known to the researchers.

Uncertainty was also generated in both computer and user departments because of the need of the firm to acquire and assimilate a new and possibly untried variety of computer technology. This uncertainty arose through lack of experience with, and knowledge of, the form of computer technology that was being considered. It arose also because of the need for both groups to collect and assimilate a great deal of new information in order to fill this knowledge gap. It arose, too, because neither group knew the best method for evaluating and ordering this information and for reducing it to an amount small enough for a considered decision to be taken. The penalties for errors in these areas were likely to be considerable in view of the amount of financial investment involved, yet the rewards of success were uncertain and unpredictable. People were being asked to embark on a high risk situation without any guarantee of eventual gains.

Poor internal communications also led to uncertainty in those firms which did not organise their decision-making and consultative groups to represent all interests. Thus a group given responsibility for 'searching' for a certain kind of information might be unsure of how other groups such as top management wished it to proceed. External communication was another uncertainty factor, for proposals made internally might prove to be unacceptable to the major outside participants in the technical decision process: the manufacturers of the computer hardware.

In AEK and Falcon Ltd it was difficult to obtain reliable information on the extent to which such uncertainties had affected the decision processes, for we arrived in both firms shortly before implementation of the new computer systems was about to begin. Our information on the earlier decision processes was therefore historical and depended on the memories of those people who had taken part. In Poultons and Grants we were able to observe for ourselves the nature and effects of some of these uncertainties and in the former we made a detailed study of the uncertainties of the search processes and how these affected the attitudes of the computer group. In Grants we concentrated our attention on uncertainties generated by the conflict of interest that developed between computer specialists and the user department. These are described in a later chapter. In this chapter we concentrate on uncertainties within the computer group deriving from the nature of the search processes they are conducting and the environment in which they are operating. This data comes mainly from our observations in Poultons.

Conflicting interests as a source of uncertainty within the computer group

In Poultons a source of uncertainty affecting interpersonal relationships and group morale arose through differences of opinion on how problems related to the choice of computer hardware should be resolved. These differences of opinion were to some extent a product of interpersonal conflict at department management level for the department had a senior manager and two second level managers whose interests did not always coincide, leading to personal differences which were reflected in the attitudes of their subordinates. Differences of opinion were, however, much more a consequence of the various groups within the department being given responsibility for different parts of the search process.

Each group was asked to obtain information on and to evaluate a particular type of equipment and procedure for the new computer system. Within the systems analyst group this division of responsibility led to those investigating one type of equipment becoming strongly identified with that equipment and determined to have it accepted as part of the final system. Such a reaction, although it puzzled the Computer Manager, was understandable. Obtaining information, testing the equipment out and evaluating its advantages and disadvantages was a lengthy process, in some cases lasting more than twelve months. If, after this period, the equipment was not approved and ordered, the group concerned believed that its work had been done for nothing. In addition to a regret over wasted time, any group looking at the equipment of a particular computer manufacturer was always on the receiving end of sales talk and of other kinds of marketing inducements. These selling strategies did have an effect on attitudes and were another reason for the group's identification with the manufacturer's product. Another factor causing uncertainty arose from the difficulty of comparing equipment. In the computer field this is not easy because the hardware of different manufacturers is rarely directly comparable. Manufacturers will have different technical solutions for the same problem. A clearcut decision based on the definite advantages of one piece of equipment over others was therefore virtually impossible to achieve.

In Poultons the first major differences in the computer group arose over the choice of computer. Poultons had had a number of computers in the past and these had been of different makes. Its present machine was a Mitchell and this had proved to be very reliable. Despite this the Computer Manager decided to be detached and objective and not to order another Mitchell machine without first evaluating the machines of other manufacturers and finding out if these had more to offer for the proposed Central Office computer system. He personally had very good relationships with Mitchell and thought privately that there was no reason to change manufacturers unless another make could be shown to have clear advantages. His Managing Director also took this point of view for he was not anxious for a situation to develop in which the company had several computers each one of a different make. However, the second level managers in the Computer Department were not so enthusiastic about Mitchell and had different choice criteria from those of the Computer Manager and the Managing Director. The Programming Manager, strongly supported by his staff, proposed that an ICM computer be bought, while the Systems Manager preferred a British Electric. It seemed that whereas the Computer Manager was placing a great deal of weight on a long standing, successful relationship with the Mitchell manufacturer, the other two managers were placing more emphasis on technical considerations.

The same problem arose with the long and difficult search for appropriate input devices to attach to the chosen machine. The Computer Manager was attracted by cathode ray tube (CRT) equipment but decided initially that it was too expensive. His Systems Manager spent a very long time investigating mark sensing document readers and came out strongly in favour of these. This research for suitable input devices then fed back on the make of computer decision and caused this to be reconsidered. At a comparatively late stage in the decision process the Computer Manager favoured a Mitchell machine but was unsure what input device could be used with it. His Systems Manager wanted British Electric and knew that this manufacturer could supply a compatible mark sensing reading device for his machine. The Programming Manager asked that ICM be reconsidered, although this manufacturer had already been eliminated as a possible supplier. It seemed that each man was reducing the felt uncertainty in the decision by appealing to his own interests.

The search process for suitable hardware was described to observers as a scientific seeking for objective information on which to base an informed choice, nonetheless it did appear to have an emotional content. The Computer Manager, who bore ultimate responsibility for the success or failure of the project, knew that if another Mitchell machine was bought his good relationships with this manufacturer would continue and he could be confident of excellent service if the machine broke down. His Systems Manager knew a great deal about the British Electric machine and its mark sensing reading device and would have had little difficulty in handling this machine if it was chosen. Perhaps, very humanly, he also wanted the pleasure of having the Board select the machine which he had spent a long time testing and evaluating. The Programming Manager had great confidence in the technical knowledge of ICM and knew that he could get first class help from this firm if he had any programming problems. Thus the choice for decision-making sent to the Board reflected to a considerable degree the personal interests of the principal investigators.

The Computer Manager later told us that he was sorry he had ever started an investigation to find the best computer for the new system. He felt that it would have been better if he had merely ordered a second Mitchell computer, for the search process had had a number of human consequences that he had not foreseen. The staff undertaking the search had to devote many months of work to investigatory activity and they suffered major swings in morale according to whether the hardware they were testing looked like being selected or not. At one point the systems analysts evaluating the British Electric machine threatened to resign if this machine was not chosen. This kind of reaction was not irrational but merely an indication that people do not like spending long periods of time on something that never bears fruit.

The search for new technical information as a source of uncertainty within the computer group

Any new area of technology is hazardous from a decision-making point of view Existing equipment is likely to be untested on any large scale or over a long period of time, while there will be little information available on future technical developments although these could affect present decisions. Computer technology is particularly prone to these difficulties. Its progress has been extremely rapid and is forecast as being even more rapid in the future. Palme (1973) suggests that mass produced integrated circuits, larger production series and new innovations will cause the computer hardware cost/performance ratio to drop sharply. He predicts that the cost for the same amount of work will reduce three to five times in every five-year period. This cost/performance factor, together with a major increase in sales, means that total available computing power will increase between nine and two hundred times in the next ten years. Thus any group selecting computer hardware at a particular time is faced with two very difficult problems: first, how to select the machine best suited for their purposes in the short term; second, how to avoid choosing equipment which will be quickly rendered obsolete by new technical developments or which in the future will prove not to be compatible with later types of computer hardware.

Poultons, as we have seen, decided to evaluate all available computer hardware before making a choice because the Computer Manager and his staff believed this to be a more scientific approach than merely buying another model of their existing computer. Their first step in the choice process was to prepare a brochure setting out the Central Office system of work and this brochure was sent to six computer manufacturers with a letter inviting them to submit quotations. A meeting was later held with these manufacturers so that they could question the firm further about its needs. Immediately there were communication difficulties for it soon became apparent that the manufacturers' representatives were unwilling to ask questions which might reveal to their rivals the kind of problem-solving approach they were proposing to adopt. Arrangements were therefore made to have private meetings with individual companies at which they could discuss, in detail, their ideas.

Further meetings were held with the computer manufacturers in the following month at which they put forward their technical proposals. By this time one manufacturer had withdrawn from the tendering process because of internal organisational problems. Four of the five remaining firms presented their ideas at lengthy meetings and left behind large, expensively bound documents for the Computer Department staff to peruse. The Poulton computer staff were now faced with the mammoth task of reading and evaluating these proposals, a task made difficult by the sheer size of the reports. Most of them had to work evenings and at weekends to do this and they believed, perhaps unjustly, that computer manufacturers operated from the premise that 'the thickest report will make the greatest impression'. Many of the reports were found to contain arithmetical errors and the systems analysts had to spend three weeks laboriously checking each arithmetical calculation concerned with the flow of data into or out of the computer.

The process of assimilating and revising the computer manufacturers' proposals caused a degree of strain to develop between different individuals in the Department. Discussions and informal meetings held amongst the system analysts became increasingly emotional as the level of work grew, particularly as the time approached when a decision would have to be taken on which computer manufacturers to place on the hardware shortlist. Five staff left during this period and work stress could have been a factor in their decision to terminate their employment with the company. As we described in the last section, systems analysts who were asked to look at the proposals of a particular manufacturer tended to identify with that manufacturer and this caused conflict to increase. But the factor generating most uncertainty at this stage was the necessity to assimilate large masses of information and to develop some criteria for evaluation which would assist its ordering and reduction.

This problem of variety reduction is one that has greatly interested cyberneticians. Beer (1972) has pointed out that high variety situations, such as the one we are describing, are hard to handle because the measure of their variety is the measure of their uncertainty. He suggests that simple mechanisms such as classifying data can greatly reduce the variety of information and make it manageable. However big the problem, its variety can in principle be halved by one decision element. He gives the example of looking for one person in a dance hall where five hundred couples are present. This presents a problem of variety of 1 : 1000, or a probability of 0·001 of making a correct selection at random. However if you can find out whether the person you are searching for is a man or a woman the problem is immediately halved. The Poulton Computer Department had this difficult task of variety reduction when choosing computer hardware.

Uncertainty generated by the large masses of information emanating from sources external to the firm was a continuing feature of the technical decision process. In addition to the hardware selection problem, there was the additional problem of choosing data input devices which would meet the firm's business needs and be compatible with the computer that was ordered. These two choice processes interacted, thus increasing uncertainty, for certain kinds of input devices could only be used with certain makes of computer.

The firm required data input equipment to enable it to transfer customer order data from the order form to the computer. Four types were available in the late 1960s. These were key punch devices used for punching data on to paper tape – this paper tape data being transferred to magnetic tape by the computer; teletype machines where the data was punched directly on to magnetic tape; reading devices with which marks or formalised script on a sensitised card were read and transferred to the computer, and visual display units on which keyed in data appeared on a cathode ray screen where it could be read and checked by the operator before transfer to the computer. More sophisticated devices which read handwriting were being developed but it was not known how long it would be before these were accurate enough to be used commercially. All these input devices had to be tested, evaluated and related to the choice of computer. Not only had they to be evaluated in terms of their existing efficiency but also in terms of future technical developments in the data input area. A shortsighted choice at this stage could mean an inability to take advantage of later innovation. For example, the firm would derive most benefit from a system by which customers' handwritten orders could be read by the computer without the need for human transcription. This kind of reader was in a developmental stage but there was no guarantee that this development would prove successful. An input system chosen because it could be easily and cheaply transferred to such a reader when it came on the market would run a risk that the device was never marketed on a commercial basis. Against this, a decision to choose another form of input device would mean that a later change to reading machines would be prohibitively expensive because it would also involve changing the make of computer.

The uncertainty associated with the accumulation of large amounts of technical information was not confined to the computer group but also affected top management who had to make the final choice of hardware and commit financial resources to it. Top management received information and recommendations from the Computer Department but, like the staff there, they had difficulty in evaluating available alternatives, in comparing these with one another and in predicting how different technical alternatives would develop in the future.

Our computer staff were therefore operating in a high risk situation, especially in terms of their relationships with senior management. Because of their technical expertise they were more competent than the Board to evaluate technical alternatives and their advice was likely to be accepted. If this advice subsequently proved to be wrong, and could be shown to be wrong because the equipment did not meet the claims that had been made for it, they risked at the worst losing their jobs, at the least losing power and influence. A major problem for them was therefore developing some mechanisms for variety reduction and information evaluation. Thompson (1967) argues that when uncertainty looms large in comparison with predictive ability, judgement is suspended and other techniques applied. These may range from the flip of a coin to quite elaborate devices for randomisation, to the use of formulas, or to precedent.

None of these approaches was thought appropriate by the Poulton computer group, or their top management, and they therefore had to develop other viable techniques of evaluation. One attempt was made to use a formula, for a firm of American consultants with an office in Britain had developed a program to compare the suitability of different makes of computers for a particular business operation. This firm was commissioned by Poultons but its finding was that any of the five makes of computer originally under consideration would be able to do the order-processing job. This conclusion was regarded by the Poulton computer team as being of little or no value. The only alternative was for them to develop their own methods. They attempted, first of all, to reduce the amount and variety of technical information through categorisation and classification. The computer manufacturers were categorised in terms of the different kinds of input devices which were compatible with their machines. The following breakdown emerged:

1 Teletype input devices   —   Universal and Mitchell
2 Keypunch   —   UDJ
3 Reading devices   —   ICM and British Electric

The next step was to reduce variety even further by evaluating the advantages and disadvantages of each of these combinations. After much discussion the systems analysts decided that the two most suitable manufacturers appeared to be ICM and British Electric, both of which used some form of reading device. Teletype input was queried because it was felt that this would require well trained operators who would be under considerable strain when operating these machines at the speed which would be required. The ICM reading device was seen as particularly advanced but as having the drawback that it was six times as expensive as the British Electric reader. It can be seen that decision-making criteria, in the form of compatibility, human constraints and financial limitations, are starting to emerge.

In another attempt to reduce variety and order the technical information, the systems analysts and programmers each ranked the available computers in order of preference, using technical criteria as the basis of their selection. The systems analysts' list was headed:

1 British Electric and ICM
2 UDJ
3 Mitchell
4 Universal

The programmers' list was:

1 British Electric and ICM
2 Mitchell
3 UDJ
4 Universal

Both these groups believed that greatest weight should be given to technical criteria. This approach conflicted with the view of the Computer Manager and Managing Director that precedent should be considered important as a decision factor. The firm already had a Mitchell computer which was operating satisfactorily and it had good relationships with the Mitchell management team. Thus a major problem at the variety reduction stage of the decision process was that groups and individuals were using different criteria as a basis for evaluation. Even when the decision criteria being used were the same, different groups and individuals were allocating different weights to the decision components.

Six months after the computer manufacturers had submitted their tenders, the Systems and Programming Managers were convinced, for technical reasons which they considered should be given a great deal of weight, that the contract should *not* be awarded to Mitchell. They gave their respective allegiances to British Electric and ICM. The Computer Manager, in contrast placed most importance upon the need for cooperative relationships with a computer manufacturer. He stressed the fact that these already existed with Mitchell.

At a later stage in the decision process when the computers had been short listed to three – ICM, British Electric and Mitchell – the Programming Manager tried to assist his own choice by listing the advantages and disadvantages of each of these machines. His evaluation was as follows:

For ICM

1 Their software is more fully developed than that of anyone else.
2 Their knowledge of computer applications and techniques is far ahead of anyone else.
3 They will be in the forefront of any future developments in computer science.

Against ICM

1 Their attitude as a company. They tend to tell everyone how to run their businesses.
2 The customer is seen as a very small fish in a very big pool.
3 They are expensive.

For British Electric

1 They are a UK Company.
2 They have a modern machine which is compatible with other machines.

Against British Electric

1 Their software is undeveloped.
2 Up to now they have only built one machine and we would not get that model anyway.
3 They have relatively little experience in the commercial field.

For Mitchell

1 The equipment exists now.
2 We know something about their software.
3 We have experience in programming a Mitchell machine.
4 We have built up contacts with them.
5 They are not expensive.

Against Mitchell

1 Mitchell is incompatible with other machines and they will bring out a new series within the next few years.

It can be seen that no one of these machines emerged as having clear advantages over the other two.

The same kind of evaluation was taking place in relation to the input devices, again without any clearly formulated criteria for evaluation and choice of either a qualitative or quantitative kind. The Computer Manager was of the opinion that a mark sensing type of reader would be backward looking and would not take account of future needs. This kind of reading machine was only able to identify simple symbols such as lines, ticks or crosses and these had to be entered on the customer's order by a clerk. Mark sensing assumed that the firm would never want to do direct transcription from documents to the computer and that there would always be some intervening stage between the original document and the computer. He now believed that direct transcription must come, as it had in banking. Therefore a teletype input with a direct line to the computer was moving in the right direction. In the interim period before direct transcription from a source document to the computer was possible, girls would punch the input on a machine.

The Programming Manager also thought that the use of mark sensing reading devices would be a retrograde step. He realised that in five years time the most advanced computer systems would be online, real time applications and that this kind of system would not fit with a mark sensing form of input. Teletype online would be a step in this direction and would allow for real time development with CRT equipment. Thus another decision criteria was now having a strong influence, namely, technological forecasts by the Poulton computer staff on how computers would develop in the future.

At this point the balance of favour lay with a Mitchell machine with the result that ICM, recognising that they were unlikely to get the computer contract, approached Poultons with an offer of a reading machine that would shortly appear on the market and was compatible with Mitchell hardware. The Computer Manager and his Programming and Systems Managers viewed this as another viable alternative, although an expensive one, for this solution would cost the firm an additional £200,000. This new option caused uncertainty to increase once again and methods had to be formulated for reducing it so as to lessen the stress of the decision-making process and assist the taking of an effective decision.

Thompson (1967) suggests that when an individual believes his resources are inadequate to cope with an uncertainty, he will seek to avoid making a decision. This was clearly a strategy open to the computer specialists for they could have asked the Board of Directors to take the decision. The three managers agreed that the problem was tremendously complex but they all felt that they must do nothing which would arouse unnecessary doubts in the minds of the directors. This, in their view, ruled out presenting a series of alternatives to the directors and asking them to make a decision. If they did this the unpredictability of the directors' response would have increased the uncertainty of the Computer Department's own situation. The Computer Manager came to the conclusion that there were two things he could do to reduce his own and his department's uncertainty. First, he could try and get some indication of the input solution favoured by the directors, thus reducing the number of alternatives for consideration. Second, with the same object in view, he could try and get Mitchell to recommend a particular input device. Unfortunately neither of these attempts to reduce variety succeeded. The directors at this time had no clearly formulated views on the subject and Mitchell refused to recommend any particular form of input. Thus the uncertainty continued and grew.

At this stage the Programming and Systems Managers, perhaps recognising that other groups were not going to help in the choice process, did a great deal to reduce uncertainty by abandoning their preferences for ICM and British Electric Computers and agreeing to the use of a Mitchell computer. This meant that the ICM reader, when developed, could be used with the Mitchell machine. In the interim period one of the other readers, such as British Electric, could be used. As a result of this new consensus the Computer Manager believed that the final decision was in sight and he began writing a report for the Board setting out the Computer Department's equipment preferences. Subject to Board approval it looked as if the uncertainty of the search process was now over, although a decision had been reached through a capitulation on the part of two of the participants in the search process and not as a result of unanimous agreement on decision criteria. Unfortunately we do not know whether this capitulation was due to an unwillingness to continue in a stalemate situation; a recognition that the balance of opinion at Board level was with a Mitchell solution, or a belief that the Mitchell machine/ICM input combination had clear technical advantages. Our hypothesis is that the second was principal factor – a realistic weighing up of the climate of opinion at Board level, together with a recognition that no other technical solution had any outstanding advantages over the one proposed. But it was to turn out that the sighs of relief over a decision being taken were premature and an outside source was soon to generate additional uncertainty for the Computer Department to contend with.

This uncertainty came from an unexpected source, namely television. The Managing Director happened to be watching a programme on computers one evening during which there was a demonstration of CRT equipment as a form of data input. Although this was expensive, and had been eliminated by the Computer Manager at an early stage in the search process for this reason, the Managing Director requested the Computer Department to look at CRT again. Thus the members of the Computer Department were faced with collecting and evaluating more information before a final decision could be taken.

These examples of the complexity of the decision-making process at Poultons show clearly the problems which innovative groups have in ordering the information which they collect and in reducing its variety. Yet unless this is done they are operating in situations of extreme uncertainty in which it is virtually impossible to arrive at a logical decision. With advanced technology, problems lie in deciding how to classify available information so as to set out alternatives. They lie also in deciding what decision criteria to use to evaluate these alternatives. A difficulty in the Poulton situation which would be found also in other firms was that the computer manufacturers all offered different kinds of computer based solution and this made direct comparisons of their proposed systems almost impossible. Comparisons on the basis of price were also seen as unsatisfactory for a low cost machine might have high running costs or might become obsolete within a short period of time.

The nature of the search process in turn generated its own uncertainty, and the fact that there was no consensus among the computer group on decision criteria meant that each new item of information collected tended to be evaluated on a different basis. We would argue that this kind of uncertainty cannot be avoided when decisions are being taken in new technical areas, with the result that this kind of decision will always be both risky and stressful. Poultons increased the uncertainty of the decision environment by deciding to start from scratch and consider all the machines that could handle the problems they wished to solve. Two of our other firms avoided this area of uncertainty by standardising and always buying their computers from the same manufacturer. Nevertheless by adopting this approach Poultons believed that they were able to take account of all available technical alternatives. They considered that through this strategy they would finish up with a better solution than would otherwise have been the case.

When a decision is being taken for a second time the learning process associated with the first attempt should assist the reduction of uncertainty. As more and more decisions of a similar kind are taken the decision process should become routinised and formal methods evolved for handling complex information inputs. In this way order is developed and uncertainty reduced. But with rapidly advancing technologies the problem is more complex for although methods for classifying and ordering information can be developed, it seems probable that methods for evaluating the advantages and disadvantages of new technical alternatives will always be complex and uncertain.

So far we have shown that two factors generating uncertainty in the computer group were, first, at the start of the search process, a lack of knowledge of the available technical alternatives; and second, as the search progressed, the mass of information that had to be assimilated and evaluated. This uncertainty could have been lessened if well tested methods for ordering and reducing this mass of information had been available, but the nature of computer technology made the development of such methods very difficult. Let us now examine in more detail some of the areas of the decision process where an inability to order and evaluate information caused particular difficulties.

First, as we have shown in the last section, there was the problem of getting agreement within the computer group itself on appropriate evaluation criteria. It seemed that individuals and groups were placing the greatest importance on those criteria which reflected their own interests. Second, there was the serious difficulty that realistic evaluations which took account of future technical developments required skill in technological forecasting, yet technological forecasting is invariably a hazardous process containing a strong element of guesswork. Third, the ordering and evaluation of data was made more difficult by the constant introduction of new information inputs by agencies located outside the company. These were the salesmen of the computer manufacturers who came along with offers of new equipment and, on occasion, with damning criticisms of the equipment of their competitors.

One way of assisting the control and evaluation of this information would have been to secure the judgment of another group. The Board of Directors could have fulfilled this role: it was they who would sign the large cheque that would eventually be required. But they were reluctant to assume it, preferring the sifting of information to be carried out by the computer group, with the results being put forward as recommendations. The Computer Department was also unwilling to use the Board in this way, fearing that an increase in uncertainty at Board level would rebound and increase their own uncertainty. The method generally used in industry for reducing the variety of information and providing a means for evaluation, namely financial measures, figured little in our case study. The reason for this related to the nature and newness of computer technology. The Computer Manager was of the opinion that the work problem solutions offered by the tendering manufacturers were all so different that financial comparisons between them would be misleading. Because of the variations in their equipment, they were all, in fact, tendering for somewhat different jobs. Also, at the time the study was carried out methods for costing computer systems were not at all well developed. This may still be the case although a number of organisations are now carrying out research into this subject. (For example, the Manchester Business School, the London School of Economics and the University of Durham who are all participating in a joint project on the economic evaluation of computer based systems.)

Uncertainty as a result of problems of internal relationships

Much uncertainty in innovative decision-taking is generated by the fact that those participating in the decision process and those who believe that they will be affected by these decisions are not sure what the end product will be. Will the new work system be a success; will it be advantageous or disadvantageous for them personally? In Poultons this kind of uncertainty was reduced in the early stages by excluding the future user departments from any role in the choice of equipment and in systems design. This approach had certain short-term benefits for the computer group although it can be argued that it increased uncertainty at the user end and probably generated more uncertainty for the Computer Department in the long term than would have been the case with earlier involvement.

The decision processes themselves, with their complexity and feedback loops, generate their own uncertainty and this uncertainty is increased by the political elements which exert an influence on most decision processes. These political factors are a product of day-to-day interpersonal relationships and of a desire by both individuals and groups either to promote their own power and influence position within the firm or to protect themselves from any erosion of existing influence or security. The unfreezing of the normal situation which is caused by change often leads to alterations in the organisational balance of power with advantages for some and disadvantages for others. In chapter 5 we describe how the nature of relationships within a computer group generated uncertainty for the group. In this section we examine how the Computer Department's relationships with the Board and with computer manufacturers can be a source of uncertainty.

In Poultons there was a considerable personal distance between the members of the Board and the Computer Department. This caused relationships to be formal and communication to be occasional rather than continual, based on formal meetings and reports. This lack of communication meant that the Computer Department tended to pursue a search process that fitted their definition of the firm's problem, without any certainty that their definition was the same as that of the Board. They were therefore taken by surprise when the Board intervened and changed the objectives of the proposed computer system. This happened twelve months after tenders were submitted by the computer manufacturers when, at the initiative of the Managing Director, the parameters of the job were widened to cover other business functions, with the result that initial goals had to be reformulated. The stimulus for this change of plan was a report submitted to the Board by the Computer Manager. The Directors told the Computer Manager that his report was a good one, but that it was directed at doing only a limited job. They had come to the conclusion that in view of the amount of upheaval that was going to be involved in introducing a new system, the Company should try to do something more ambitious.

A number of external influences appeared to have influenced the Directors' rethinking of system objectives. One was the fact that the firm's major competitor had announced in the press that it was embarking on a massive on-line system using CRT terminals. Another was that a large government department with a high demand for clerical labour had opened new offices close to Poultons and this was likely to attract away some of the firm's clerical labour force, already difficult to recruit. It was at this time that the Managing Director asked the Computer Manager if he would reinvestigate CRT equipment for input. The Directors asked him to obtain quotes for CRT equipment in relation to the new, larger computer operation they were proposing.

One result of these changed objectives was that the Computer Manager had a meeting with his Programming and Systems Managers. As a group they were faced with having to answer a number of new questions put to them by top management. The first of these was 'Is CRT input feasible?' The second, How do we restructure the Computer Department's objectives now that we are working on a much bigger computer system?' Although Mitchell had now been accepted by all of them as the manufacturer to provide the computer, the Programming Manager suggested that this decision should be reconsidered. He felt that as the project had become so large and ambitious another manufacturer might be preferable. ICM, for example, had great experience with this kind of system and its enormous research expenditure must mean that it would stay ahead of other computer manufacturers. The Programming Manager valued ICM's software expertise for, from his point of view, software was the most difficult aspect of implementing a new data-processing system. Thus their relationships with the Poulton Board generated uncertainty for the Computer Department, especially the fact that they were not always *au fait* with the Board's latest thinking. Until a late stage in the search process the Computer Department were never entirely convinced that the directors were truly committed to the proposed computer system. No one would have been surprised if the Board had decided to cancel the project, and this fear was an additional source of uncertainty.

Relationships with the computer manufacturers added to the uncertainty with which the computer staff had to contend. Manufacturer's representatives constantly intervened in the decision processes in an attempt to discredit their competitors or demonstrate that their equipment was superior. For example, a manufacturer whose machine had been eliminated at an early stage visited the firm and expressed great astonishment at the nature of the short list, pointing out that one of the listed machines had very unsatisfactory software. This intervention had a considerable impact on the Computer Manager who wondered if his firm had, as one of its possible choices, a machine that could not do the job. This particular machine was a new one and there were few examples of it in use in Britain. A second manufacturer circulated rumours that a manufacturer on the short list was about to bring out a new range of computers which would not be compatible with their existing machines. If correct, Poulton's might buy a machine that could not be used with later models.

Some months later another of the rejected computer manufacturers tried to get back into the picture. He visited the Computer Manager and told him that his company was bringing a new machine onto the market which would be available in two years time. He suggested that his Company should rent Poultons one of their existing machines on a yearly basis until the new machine was available. After consideration the Computer Manager dismissed this idea as impractical. A further intervention came from ICM. When the area of business to be covered by computer system was extended the Programming Manager suggested that the make of computer should be reconsidered and that ICM might be preferable to Mitchell. Within a week of this remark the ICM sales representative visited Poultons and offered to organise a seminar for top management which would provide information on his firm's machines. Even when the contract with Mitchell was finally signed one computer manufacturer wrote to the Board of Directors stressing what a serious mistake they had made in their choice of hardware.

In this way the competitive marketing activities of the various computer manufacturers did a great deal to increase the uncertainty of the decision-making process. We felt that the computer market must be one of the few areas of selling where a rejected vendor is prepared to query the buyers' choice of a competitor's product. Throughout the whole of the search process manufacturers excluded from the short list made determined efforts to be reconsidered. Those on the short list made attempts to influence different members of the Computer Department through offers of hospitality and much wining and dining. Although Computer Department staff recognised all these overtures as good salesmanship they nonetheless added yet another complication to the Department's investigations.

How our firms coped with this kind of uncertainty

Rappoport (1967) tells us that 'freedom from uncertainty is a luxury rarely enjoyed by the manager. . . . Notwithstanding the difficulties the decision maker is expected to attain satisfactory results as measured by the goals of the organization.' All our firms encountered similar kinds of uncertainty while carrying out the long and expensive search processes they believed to be necessary before arriving at final decisions on their computer systems. It was not always easy for them to know when they had identified a suitable procedure or piece of equipment for solving a particular problem and should therefore stop their investigations. We have seen that in one firm the Computer Manager admitted that his personal desire to approach the solution of his firm's problem in a detached and scientific manner had greatly prolonged the search processes and increased the uncertainty of the decision-making situation. In retrospect he believed that it would have been simpler and more effective to have implemented the solution that he felt to be the best when the investigation began. In other words he now believed that a satisficing solution would have been preferable in terms of time, cost and stress to the optimal solution which he was seeking. On this point Rappoport tells us that 'a theory of the cost and value of information for information decisions is critically needed. In an increasingly technological society, information costs are ascending to previously unimagined heights.' We would argue that a hidden cost factor is the uncertainty generated in the decision process by an extended search process and by the stressful effect this can have on the people involved in it.

Another factor which tended to increase uncertainty and prolong all search processes was the reformulation of goals by top management as projects developed. These altered goals emerged as a result of new technological opportunities; they also emerged because the length of the search process gave top management plenty of time to reformulate and restructure the problems it thought should be solved. An additional reason for a change or modification of goals was the fact that at the beginning of a project top management generally set goals of a strategic nature. That is, they indicated the kind of situation which they hoped to achieve as a result of introducing new computer systems. These broad strategic goals then had to be operationalised and broken down into a number of subgoals and it was these subgoals which tended to change as top management recognised that there were other more desirable ways of achieving their strategic objectives.

Each firm developed its own methods for coping with the uncertainty of the search process. Within AEK, goal setting and problem identification proceeded via the mechanism of working parties and committees which were individually given responsibility for different kinds of information search or different areas of information evaluation. As new possibilities for improving problem solution appeared, so additional committees were set up to evaluate these. At the time of our study the Treasury had laid down a procedure for EDP projects which instructed government departments to go through the following search and decision stages.

1 a preliminary study in broad terms;
2 development of a model EDP system based on a hypothetical computer;
3 a formal invitation to computer manufacturers to tender;
4 preparation for delivery;
5 parallel running and takeover;

The model EDP system was developed through a large and detailed fact-finding exercise in the areas being considered for computerisation. This was a job specification exercise from which a model system could be developed. The final report set out the tasks to be assisted by the new computer system, the features of the proposed system and a future policy for the system. Only after this report was approved by the Treasury were computer manufacturers asked to submit tenders. Falcon Ltd adopted similar, though somewhat less formalised, procedures, while, as we have seen, Poultons preferred to leave most of the work to its computer staff, with top management making periodic evaluations of their reports.

Clearly, each organisation had developed mechanisms for collecting, sifting and evaluating information which best suited their particular needs and cultures. The method used by AEK was similar to a management science approach with data being collected and developed into models, although of a non-mathematical kind, before action was taken. It came closer to an algorithmic method of decision-making than the approach used by Poultons which was more heuristic and closer to an adaptive model of planning and decision-making.

All our firms tended to concentrate their search processes on technical

factors and the development of technical solutions. They paid less attention
to human factors and human solutions, though sociologists would argue that
these are equally important in the operational success of the final system and
therefore should be given equal weight throughout the decision process. Here
we enter the realm of 'values' and we found that the values of the participants
in the decision processes were predominantly technical. They did not have
the social engineering knowledge which would enable them to take more
account of human factors.

The most striking feature of the decision processes which we observed was
the length and comprehensiveness of the search for information. A major
problem for the information gathering group was knowing when to stop
searching. Poultons were reluctant to stop because they recognised that any
solution, once accepted and implemented, immediately became obsolete as
new technical opportunities appeared on the scene. This attitude was somewhat at odds with the attitudes and objectives of their Board of Directors.
The directors were looking for good computer based solutions to solve
immediate business problems. The computer group, in contrast, had a
powerful innovatory value system and were motivated to achieve a high level
technical solution.

To sum up, the way the search for information was handled was a product
of the manner in which preliminary goals were defined. We have seen that
these started as broad, strategic goals with a technical rather than a human
content. As information was collected new opportunities were recognised,
either for solving more successfully the problem as originally defined, or for
coping effectively with an extended problem area. These new opportunities
appeared both as a result of the investigations of the systems analysts and
through the information inputs of external agents such as computer manufacturers. In this way the search for knowledge took an interactive form with
solutions looking for problems as well as problems looking for solutions. This
boundary-spanning activity had its own impact on the uncertainty of the
decision-making situation. The more information that was collected, the more
difficult it became to arrive at a final decision, although any final decision
based on more information was likely to be superior to one taken on less information. Negative interventions from outside by computer salesmen who
wished to discredit the products of their competitors increased uncertainty as
the computer group had difficulty in knowing how much attention they
should pay to these. The Poulton Computer Manager tried to protect himself
from this kind of uncertainty by only seeing Mitchell computer salesmen; he
referred the salesmen of other computer manufacturers to his managers.

Uncertainty can be reduced by a willingness to satisfice rather than optimise and by examining and evaluating only those alternatives which are con-
spicuous. But because the level of financial investment in these computer
systems was so great and the possibility of failure so damaging, all our firms
undertook a deep and most exacting level of search. They therefore consciously increased the uncertainty of their decision-making environments,

believing that by doing so they were reducing the risk of implementing an unsatisfactory system.

If a group is taking decisions in an open system environment – that is, an environment in which departmental and organisational boundaries are crossed and there is considerable interaction with groups in other parts of the firm, or outside the firm – then attention has to be given to developing means for reconciling different interests and taking account of different philosophies and values. All our computer groups had to negotiate with top management and user departments within their own firm. They had also to negotiate with computer manufacturers outside the firm and, if their firm was part of a larger organisation, with central and head office groups. To cope with possible conflicts of interest, mediating and conflict-resolving mechanisms were required for situations where the reconciliation of different attitudes and values was likely to prove difficult. There had to be a recognition that most innovative decision-making occurs in a highly charged political environment. It does not resemble the rational choice situation described in some of the early theories of decision-making.

Mediating and conflict-resolving mechanisms are an important means for reducing the kind of uncertainty which is a product of interpersonal relationships, and each of our firms approached this problem in a rather different way. In AEK the formal committee structure for information evaluation and decision-making also acted as a means for clarifying the nature of different interests and for resolving conflict. The most outstanding feature of the decision process in AEK was its democratic nature and the care with which all interested groups were involved and all recommendations carefully checked. This ensured that strategic objectives and subgoals were accepted by all parties to the decision; that problems were identified and defined in terms of all interests, and that the decision process incorporated checking mechanisms to ensure that a sound decision had been taken and agreed to before the next stage of the search process was begun. This approach, although somewhat slow and unwieldy, did appear to result in AEK being less affected by major uncertainties during the decision process than was the case in the other firms. The fact that all interested groups had clearly defined and well understood search and decision responsibilities seemed to reduce the amount of internal politicking and attempts to influence the decision to suit particular vested interests. Falcon Ltd followed a similar, though less precisely structured approach. Plans were carefully formulated and tested out before implementation, and as one computer system was being introduced long-term planning for subsequent systems was taking place.

Our two other firms would not have wished to have adopted such a bureaucratic approach, as the computer groups would have seen such a large committee structure as inhibiting their own freedom of action. They wished to achieve a solution which met their own high technical standards as well as the business interests of the firm. Nonetheless, because they operated in a more fluid way and with more autonomy, they did increase the amount of uncertainty they had to face. The exclusion of user departments from early stages of the decision processes rebounded by making relations difficult at a later stage. In Poultons decisions were also made difficult because of communication gaps between computer specialists and top management. We felt that the isolation of the computer group in Poultons from both users and top management led them to develop vested interests in particular solutions. This, in turn, led to politicking within the Computer Department similar to that which we describe in the next chapter.

5

Internal politics as a source of uncertainty

A factor which can generate considerable uncertainty when innovative decisions are being taken is the nature of the internal political processes associated with decision-making activity. Participants are likely to view the organisational changes which will be a consequence of the decisions either as opportunities for reinforcing their own position in the firm through gains in power and influence or as threats which may lead to a decrease in influence, status and even job security. When perceptions are of this kind the decision process will be seen as a battleground in which the prizes are the increase or maintenance of power, influence, status and security.

Surprisingly, little attention has been paid to the political aspects of decision-making, and the extent to which the internal political environment is itself a major source of uncertainty when an important innovation is being introduced (an exception is Pettigrew (1973b)). There appear to be several reasons for this neglect. First, while sociologists have always been interested in the structure of organisations they have tended to confine their analysis to the organisation as an authority structure with formal hierarchies, rules and regulations and to neglect the organisation as a political structure – that is, as made up of groups of people who are seeking to exert influence on formal and informal strategies related to the allocation of resources. Secondly, information on political behaviour is not easy for a research worker to obtain. As Tom Burns (1962) has pointed out,

> the problem is no one regards himself as a politician, or as acting politically, except of course on occasions when he is led into accounts of successful intrigue and manoeuvering when he bolsters up his self esteem and reputation by projecting the whole affair in the context of a game or joke.

There is the additional problem that those who are politically involved usually claim that they are acting in the interests of the company as a whole; this is how they legitimate their behaviour. Nevertheless a few studies of political behaviour in industry have been carried out and published. The best known are those by Dalton (1959) who describes the informal power relationships affecting the behaviour of production and maintenance departments and line and staff personnel and Crozier (1964). Crozier shows how people low down in the hierarchy can control the behaviour of those above them because they are in charge of an area where there is a degree of novelty and uncertainty. This inhibits tight control from above and makes the supervisor dependent upon the knowledge of his subordinate.

Although politics as a factor in decision-making behaviour is given little attention, one of its manifestations, conflict, is discussed in the literature. In chapter 3 we showed how March and Simon (1958) view organisations as choosing, decision-making organisms capable of doing only one or a few things at a time, and of attending to only a small part of the information recorded in their memory and presented by the environment. March and Simon believe that the interdependent nature of the different parts of an organisation produces a 'felt need for joint decision-making'. They see conflict occurring in the decision-making process if there is a difference in the goals of the various participating groups and if there is a difference in the knowledge and perceptions of these groups. For March and Simon conflict is an abhorrent and undesirable phenomenon and they suggest that once it is recognised steps will be taken to resolve it.

Cyert and March (1963) take these ideas a stage further by stressing the concept of the organisation as a coalition. In contrast to the earlier thinking of March and Simon (1958), they regard conflicts of interest which are a result of different group goals as 'normal' parts of organisational life. They discuss the mechanisms of conflict resolution but accept that such conflict resolution can only be partially successful, pointing out, 'most organisations, most of the time, exist and thrive with considerable latent conflict of goals'.

Cyert, March and Simon have made major contributions to our understanding of organisational decision-making but later research now suggests areas where their ideas can be developed even further. It can be argued that this earlier work on decision theory places too much emphasis on how the individual affects the organisation, and does not take sufficient account of how organisational pressures and constraints influence the behaviour of the individual. This leads March and Simon to pay a great deal of attention to the individual as an information-processing system and not enough to considering the organisation in the same terms. Their opinion that conflict is due only to differences in group goals and to differences in the knowledge and perspective of interacting groups implies that once these differences are recognised they can be resolved without too much difficulty and that groups will make every effort to do this. It can be argued that there may be other reasons for conflict when decisions are being made. For example, one group may see the result of a decision route taking a particular direction as leading to an enhancement of their own organisational power position and the diminution of a rival group's power. This may make them push their own interests and, in turn, will evoke opposition in the rival group. Thus the decision-making process may be used for the furtherance of conflicts which are a product of long-standing organisational rivalries. If this happens then battles for power become a major source of uncertainty.

Cyert and March (1963) recognise the importance of the formation of coalitions, and perceive the necessity for subgroups to obtain support for their particular interests but they do not give detailed attention to the processes by which this support is generated. They are well aware of the influence of individual or group power on decision-making, stating, 'any alternative that satisfies the constraints and secures suitably important support within the organisation is likely to be adopted.' But they do not explain how such support is obtained by those interested in securing the acceptance of a particular problem solution, nor do they consider the fact that the way an organisation is structured may limit a group's ability to secure support. This leads to gaps in their decision theory as it is not possible to explain why particular alternatives are raised at specific times in the decision-making process, by whom and with what consequences.

In this chapter we argue that political behaviour is likely to be a particular feature of large-scale innovative decisions such as those associated with the introduction of computer systems, and that it will be an important contributing factor to the uncertainty confronting the decision makers and those affected by the decisions. Innovative decisions generate uncertainty because they have the potential to alter existing patterns of resource sharing. New resources may be created by an innovation and come within the jurisdiction of a department, group or individual which has not previously laid claim to this form of resource. This department, group or individual may then perceive these resources as an opportunity for gaining status and power relative to others in the organisation. At the same time those who see their interests as threatened by the proposed change may resist such an allocation of new resources and this will influence their behaviour in the decision-making process. In this way political action is generated. Some of the earliest observers of the effect of change on the power structure of an organisation were Burns and Stalker (1961). Describing the electronics industry in Scotland after the last war, Burns writes:

> As production and the market have moved into a fundamentally unstable relationship, and as the stream of technical innovation has quickened, the legitimacy of the hierarchical pyramid of management bureaucracy has been threatened by the sheer volume of novel tasks and problems confronting industrial concerns. . . . New and unfamiliar tasks and problems create situations and demands for information and action to meet them which are often incompatible with the presumption of the traditional hierarchy.

The reasons for political behaviour

Resources which give an individual or group power and influence and which may be redistributed as a result of organisational change will include the ability to provide desired services, specialised knowledge, informal influence positions – such as access to the ear of the boss because you meet him socially – and hierarchical authority. They will give A power over B to the extent that, using them, A can get B to do something that B would not otherwise do (Dahl, 1957). Power then becomes a property of social relationships rather than something which an individual has as a personal asset. This theory of power implies dependency as a part of these social relationships. Blau (1964) suggests that, 'by supplying services in demand to others, a person establishes power over them. If he regularly renders needed services they cannot readily obtain elsewhere, others become dependent on and obligated to him for these services.'

Thus the power of one individual over another is related to the importance and uniqueness of the services provided by the first individual and the alternative means for obtaining these same services open to the second individual. In the later sections of this chapter we see how these ideas of power and influence affect the politics of innovative decisions.

The theory which we present below explains aspects of decision-making behaviour in terms of political factors. The organisation is viewed as an open political system containing subunits which have developed particular interests as a result of their specialised functions and responsibilities. But although each subunit has its own set of interests and its own specialised tasks it has to relate with many other subunits and to work interdependently with these. Both this specialism and this interdependence affects any decision-making process and as part of this process interest-based demands will be made. In the absence of any agreed set of priorities in these demands, conflict is likely to ensue with subgroups competing for scarce resources in order to promote their own interests. These resources may be of many kinds and will frequently be in short supply. Capital expenditure is one example. A group may seek to increase its share of the firm's financial resources, as this will enable it to extend its work through the purchase of new equipment, or increase its area of influence through the hiring of more specialist staff. Other desired resources will be control over people, information or new areas of a business.

The extent to which a particular claim is prosecuted will depend on how critical this factor appears in terms of the survival and development of the group, or of key individuals within the group. The success of groups and individuals in furthering their interests will be a consequence of their ability to generate support for their demands. We would stress the importance of individual goals in decision-making behaviour. It is not always recognised that the desire of individuals to increase or maintain their power position may have as much influence on the decision-making process as departmental or other subunit goals. Soelberg (1963) suggests that 'the desire for power and concern for personal advancement represent goals which are of central concern to an organisational theory of decision making'.

Politics then can be seen as one of the mechanisms by which individuals and groups seek to obtain power over others. They do this in order to secure some advantages which they believe will assist the achievement of personal and group goals.

Burns (1965) has developed a theory to explain this political behaviour:

> The notion either of a hierarchy of sub-goals which, although generated within, and by the existence of, the organisation, wander out of line so far as organisational goals are concerned, or of an organisational goal generated by the concensus reached by individuals, each with personal goals, bargaining and learning their way towards a satisfactory equilibrium between their goals and those of the working community can itself only be realised and made operational if we accept the fact that the organisation represents only one of several means – and systems for realising the goals of the individual.

Burns emphasises that it is important to recognise the fact that the individual has personal as well as organisational commitments. He calls these political and career commitments. The aim of political commitments is to increase the individual's personal power through his attachment to groups of people associated with the kind of resource he seeks who are interested in enhancing its value, or to cabals seeking to control or influence the exercise of power. Positions of power appear to offer their occupants both material and psychological rewards and this in itself stimulates conflict between those who have power and those who want it. This desire for power can bring into play a multiplicity of strategies of coercion, deceit and manipulation which can be used either to acquire, or to hold on to power.

What we are saying is that within decision-making processes power strategies are made through demands by the various interested parties. Strategies have been defined as 'the links between the intentions and perceptions of officials and the political system that imposes restraints and creates opportunities for them' (Wildarsky, 1964). A demand 'is an expression of opinion that an authoritative allocation with regard to a particular subject matter should or should not be made by those responsible for doing so' (Easton, 1965). A distinction can be made between an interest and a demand. The expression of an interest is not identical with the input of a demand; for an interest to become a demand a proposal for authoritative action to be taken is required.

The more complex heterogeneous and differentiated a political structure the more likely are disparate demands to be made. Such disparities are a product of the uncertainty and complexity of the task in hand, organisational position, professional training, group norms and values, pressure from external groups with an interest in the decision outcome, and the history of relationships and attitudes between the groups which are making demands. If there is no clearly set system of priorities between demands there will be conflict. The way demands are dealt with by the groups involved and the generation of support by particular groups for their own demands are the principle components of the general political structure through which power is wielded. The

success of an individual or group in achieving its demands will be a consequence of ability to generate support. The final decision will be greatly
influenced by the processes of power mobilisation attempted by each party
to the decision-making process in support of its own demand.

The involvement of individuals and groups in such demand and support
generating processes within the organisation constitutes the political dimension of decision-making activity. Political behaviour is defined as behaviour
by individuals, or groups, which makes a claim for organisational resources.

If the analysis set out above is accepted, this leads us away from a concentration on notions of satisficing as an explanation of decision-making behaviour to alternative explanations of choice among alternatives as a product of
the strategic mobilisation of power resources. These resources must not only
be possessed by the power aspirant, they must also be controlled by him.
Bannester (1969) makes this point when he says: 'It is immaterial who owns
the gun and is licensed to carry it; the question is, who has his finger on the
trigger?' An undue emphasis on political considerations may be a feature of
major innovative decisions simply because it is extremely difficult to predict
the consequences of one particular technical choice against another, or even
to know when a satisfactory solution has been put forward.

Frischmuth and Allen (1968) describe this problem:

> For a technical problem there is no correct, or even best, solution in the
> long run. In fact, there is frequently no terminal state; the solutions them-
> selves are often dynamic. The interaction of the researchers with their
> environment is also continual and changing. The goals for the problem's
> solution may be established when the process is initiated, but they are
> subject to change as the process proceeds. They do not explicitly contain
> the criteria by which the solution is to be evaluated, since the criteria are a
> matter of judgement and differ among individual evaluators.

To sum up the ideas put forward so far we suggest that individuals within
an organisation are likely to use decision-making situations to pursue what
they perceive to be their own interests or the interests of the groups to which
they are affiliated. The extent of their ability to pursue these interests will be
a product of their power position and this in turn, will depend on their ability
to influence the attitudes and behaviour of others through their possession of
some scarce resource. This scarce resource, which can be information, skill
or knowledge, will be seen by these others as critical to the successful choice
of a solution for the decision problem. The fact that this kind of political
activity takes place adds to the uncertainty of decision-making.

It must be recognised that so long as organisations continue as resource-
sharing systems, where there is a scarcity of desired resources political
behaviour will occur. This does not mean there is such a finite distribution of
organisational resources that the giving of something to Peter inevitably re-
moves something from Paul. It is quite possible for resources as a whole to
be increased. But allocation is likely to be a problem everywhere and the
power distribution within an organisation at any one moment will be an

important factor in determining who shall gain a disproportionate share in new resources as they become available.

We have placed a great deal of emphasis on the suggestion that individuals and groups make choices and influence others to make choices which are in line with their own personal interests. But it must be recognised that not all individuals act in line with their own self-interests. Blau (1964), for example, points out that not all human behaviour is dominated by an individual's pursuit of social rewards. Nevertheless it is difficult to conceive of much behaviour within a profit orientated commercial situation which does not have an element of self-interest associated with it, and the arguments put forward in this chapter are based on the hypothesis suggested by Blau (1964) 'that men seek to adjust social conditions to achieve their ends'.

Factors influencing political behaviour

In the last section we argued that man seeks to control personal uncertainty and to stabilise or reinforce his power and influence position through manipulating fluid situations to fit his own interests. Large-scale innovative decisions such as the introduction of big computer systems are prone to be affected in this way for one of their consequences can be a redistribution of scarce resources such as power and status within the organisation. This pursuit of personal or group interest generates uncertainty for the decision process as a whole, for the participants and for those who will be affected by the eventual decisions.

Let us now briefly examine some factors in the environment of the individual or group which reinforce and direct this protection of self-interest within the decision-making situation. The first of these is the nature and position of the role which the individual or group occupies within the organisation structure and the manner in which this differs from other roles. Sociologists call this difference 'structural differentiation'. In this chapter we consider briefly the role of the 'expert'. In chapter 6 we look in more detail at how occupational roles develop over time and at the factors which influence role definition.

The role of the expert

An individual occupies a certain role within an organisational structure and is responsible for a certain set of tasks; nevertheless in the pursuance of organisational objectives he has to interact with other individuals and groups whose responsibilities are different from his own. If this did not happen the organisation would be unable to produce the goods and services which are its primary function. This differentiation between work roles is a product of many different structural features but is broadly based on differences of technology, territory and time. Lawrence and Lorsch (1967) for example, split the firm into three subsystems: production, sales and research and development and suggest that each of these subsystems varies in its structure, goals and time constraints; these variations being a product of differences in the environmental uncertainty of each subsystem. For example, production has to respond to changes in product markets with new types of goods; sales must seek to find new markets which will permit an expansion in production or a compensation for business that has been lost. R and D must develop new products which will also assist sales.

A particular feature of innovative decisions is the role of the technical expert. He is the individual who has the specialised knowledge to search out and evaluate the technical information which is crucial to the decision-making situation and the eventual success of the new system. Mechanic (1962) has argued that within organisations dependency relations are created through groups or individuals controlling the access to resources such as 'information, persons and instrumentalities'. Clearly control over information is a critical factor in the ability to influence choice in decision-making situations. In innovative decisions concerning the use of computers, where the decision-making body normally includes both line executives and computer experts, the power of the experts over the line executives is likely to be related to the dependence of these individuals and groups on the technical knowledge of the experts. Experts can as a rule maintain a power position over high-ranking persons in an organisation so long as the latter are dependent on them for special skills and access to strategic information. Experts are also given power if they are perceived by senior management as facilitating the ability of the organisation to cope with uncertainties and pressures originating in its environment (Hickson *et al.*, 1971). This suggests that the power of the expert will be greatest when he is taking innovative decisions in an area critical to the survival of the organisation.

We have already discussed the fact that innovative decisions are characterised by uncertainty, and this uncertainty can be used as a major power source by the expert. Crozier (1964) provides an example of a technical engineer who was able to control the actions of his director by setting technical limits on what it was possible and what it was not possible to do. But the power of the expert in a joint decision process is unlikely to be omnipotent even with the most technically uncertain problem, for the position the expert occupies in the structure of relations in the organisation will certainly affect his ability to control and direct the actions of others. Senior executives generally have ultimate power to hire and fire experts, and this fact alone exerts a major control over expert strategies. Experts who do not sit on a company board are rarely sufficiently strong to get technical strategies accepted which are opposed by powerful interests at the top of the organisation.

The expert may not rely for his influence solely on the power his unique knowledge and access to information gives him. He will also seek to generate support for the demands he is making and the strategies he is formulating, and the amount of support he is able to secure from other individuals and groups in the firm will be a consequence of the nature and structure of his organisational relationships. The personal respect in which he is held may be an important factor, as will be his general personal acceptability and the extent that others feel indebted to him for past services. Even the timing and manner of the way in which he seeks support may have crucial impact on the success he achieves (Devons, 1950). In short, 'where support is lacking, it may be mobilised; where attention is unfocused, it may be directed by advertising; where merits are not obvious, they may be presented in striking form' (Wildavsky, 1964).

Crozier (1964) has analysed the evolution of power relationships within organisations and noted the self-defeating nature of expert power. So long as the expert is operating in a new area requiring novel and unprogrammed decisions his knowledge will give him power. But as soon as a new field of knowledge becomes well covered and first intuitions and innovations become translated into rules and programmes, the expert's power disappears. The expert, for his part, is influenced by the nature of his specialisation and the direction in which this biases his search and evaluation procedures. He is influenced and constrained by his position in the organisational hierarchy and the extent to which he has access to other, more powerful groups such as the board of directors. He is also influenced by the nature and extent of his external affiliations – for example, pressure from outside salesmen or membership of professional groups.

Historical influences on relationships

Group relationships are never static, but always in a state of development. Every group has a history and present relationships are a result of earlier experiences which members of the group have gone through together. These experiences may be a result of changes in the environment of the group, and the need to adapt to these can bring added cohesion; or they may come about through the changing patterns of relationship in the group as coalitions are formed and disbanded and individuals form or break friendship links. To achieve any real understanding of the dynamics of political behaviour it is necessary to take a historical perspective and to identify and undertsand those factors in the group's background which are exerting an influence on current perceptions, interests and behaviour. In the section on 'manifestations of political behaviour' below we give some examples of how the history of relationships in one computer department greatly influenced the behaviour of its members when they were taking an innovative decision.

Complexity, uncertainty and risk

Complexity, uncertainty and risk are features of organisational decision-making with which most organisations must live. Many writers have discussed the significance of this. Devons (1950) has talked of the 'inevitable uncertainties of planning in conditions of rapid technical development'. Downs (1967) argues that 'important aspects of many problems involve information that cannot be procured at all, especially concerning future events; hence many decisions are made in the face of some eradicable uncertainty'. Lawrence and Lorsch (1967) have defined uncertainty as consisting of three components:

1 the lack of clarity of information;
2 the long timespan of definitive feedback;
3 the general uncertainty of causal relationships.

Uncertainty appears to have two dimensions: the simple-complex, and the static-dynamic (Dill, 1958; Thompson, 1967; Duncan, 1970). The simple-complex dimension relates to the number and variety of factors in a decision unit's environment which impinge on the decision unit's behaviour. The static-dynamic dimension relates to the extent to which factors in the decision unit's environment are stable or in a continual state of change over time. The case study firm examined under the heading 'Manifestations of political behaviour' clearly comes at the end of these continuums, with a large number of variables to be taken into account by the decision-making groups and with a high degree of dynamism in the decision situation as objectives were re-formulated and new technical possibilities became available.

External forces which influence the decision process

Just as no man is an island, no group or organisation in modern society is completely isolated and internal behaviour will be greatly influenced by pressures and other forms of stimuli impacting on the group from its environment. With large-scale technical decisions involving the purchase of expensive hardware, the outside manufacturers of this hardware will have a major incentive to influence the internal decision process and their marketing strategies are likely to contribute to the general uncertainty of the decision situation. It can be argued that one of the objectives of the sales staff of an equipment manufacturer is to generate uncertainty, particularly in relation to the proposals of their rivals. In addition equipment manufacturers are selling to many companies at one and the same time. This means that they are likely to have to distribute their sales resources amongst a number of customers and, because of this their marketing interventions are likely to be sporadic and therefore unsettling to a group attempting to come to a clear and rational decision on what hardware to buy. A good salesman needs to understand the political factors within a firm which are influencing its technical decisions and to intervene in these in such a way that his firm's interests are favoured. Because of this need the effective salesman will spend a great deal of time considering the power structure of the customer firm and identifying which individuals or groups will carry the most weight in the final decision choice. These attempts to interfere with and influence internal political processes will, in turn, add to the uncertainty of the internal decision-making

environment and may increase political activity within those groups responsible for taking the final decision.

The factors we have described – role, specialisation, position in the organisation structure and external affiliations – will all affect the amount of power
an individual or group can wield to support personal interests. They will also
influence the kinds of alternatives that are examined by the decision-making
group, the timing of their appearance, how they are justified and attacked, by
whom, and ultimately which alternative is chosen. They are thus potent
variables in the politics of the decision-making situation. Other factors which
generate and affect political behaviour are the history of relationships within
the decision-making group, the complexity, uncertainty and risk associated
with large-scale innovative decisions and the pressure exerted by external
groups with strong vested interests in the decision outcome.

Manifestations of political behaviour

We have put forward our hypotheses on the reasons for political behaviour:
attempts to protect or enhance individual or group interests through securing
an increased share of scarce resources such as power, influence and status, or
through preventing an erosion of these. We have argued that the factors discussed in the last section will affect political behaviour within the decision-
making situation and that personal or group power will have a major influence
on decision processes. This power being allocated to an individual or group
because of an ability to assist the organisation to cope better with environmental uncertainty.

In this section we examine the form that political behaviour takes, drawing
our examples from one of our case study firms. We argue that individuals and
groups assess the decision-making situations in which they are involved and
make choices concerning decision consequences in terms of personal and
group interests. They then seek to influence other participants in the decision
process to support or accept these choices. In order to exert this influence they
develop and use strategies which they believe will generate support or reduce
opposition. These strategies will be more or less successful depending on
the behaviour of the other participants; their perception of their interests
and their willingness to respond to the strategies. The success of the
original choices in protecting or furthering group interests will also depend on how accurately the relationship between choice and interest was
identified.

The decision situation which we describe here was common to all our case
study firms. A computer system had to be developed for an area of the
business which had previously not been computerised and this involved the
choice and purchase of a new machine and data input devices. This choice
process involved the collection of a great deal of technical information, the
evaluation of this information, the arriving at decisions on the best machine
and input devices for the firm's purposes and the presentation of these re-

commendations to the Board of Directors. The decision-making activity took twenty months to complete.

The decision participants were the Computer Manager, his Systems and Programming Managers, the Board of Directors and the external groups of computer equipment manufacturers. Line management of the future user department were not brought into the decision process until close to implementation of the new system. We describe how the various participants to the decision process viewed their individual and group interests and the choices they made on what decision factors to support or reject in order to further these interests. We are not suggesting that choices were ever made which put any of the firm's interests in jeopardy, but that because there were a number of routes available for achieving the firm's interests individuals and groups chose the one that best fitted with their own. We examine the strategies used to gain support for an individual or group's choice position and consider whether these strategies were effective. We also consider whether the nature of self-interest was always identified clearly and whether the choices that were made really assisted the furtherance of self-interest.

Members of the Computer Department had a general interest in ensuring that the firm did introduce the new system. Such an innovation meant that their specialist skills would be used and recognised as important; they would also have the excitement and challenge of developing and implementing the new system and they hoped for the eventual satisfaction of seeing it in operation and receiving the congratulations of users and top management on its efficiency and benefits. There were however some conflicts of interest between the systems and the programming group with each group seeing a particular hardware and input choice as assisting their interests more than another. There was also considerable rivalry between the Computer Manager and his Systems and Programming Managers. Each sought to increase his personal influence and power with the Board of Directors, while protecting other interests such as the ability to use his skills and knowledge in the way he considered best. This last implied the acquisition of computer hardware which he understood and could use effectively.

Political activity in the computer department centred on the make of computer that should be chosen and the kind of input devices that could best be linked to this machine. From the start of the decision process each of the three managers and both the programming and systems groups proposed different solutions to these choice problems and a great deal of political activity was used by individuals and groups to get their particular choices accepted by everyone. This political activity had a strong emotional content and led to swings in morale within the Computer Department as the power to influence the final decision moved from individual to individual and from group to group. The three managers each became personally identified with a particular make of computer and with particular input devices for reasons which were entirely rational. The Computer Manager wanted to buy another model of the make of computer which the firm already possessed. He had developed a very good relationship with this manufacturer and had confidence that if ever there were technical problems the manufacturer would provide immediate assistance. The Programming Manager and his staff were influenced by the ease with which a machine could be programmed. Their competence would be assessed through the success of the programs. The Systems Manager was an expert in a certain kind of input device and therefore wanted this device to be purchased, together with a computer which was compatible with it. Thus each individual favoured an alternative that fitted in with his own area of knowledge and eased his own task. If an equipment choice were made in terms of his preferences this would make his personal situation more secure and increase his prestige and influence within the company.

In this firm there were two phases of political activity. An early phase in which the group gradually arrived at a consensus on what hardware recommendations should be made to the Board, and a second phase, which generated a great deal of uncertainty for the group, when the Board unexpectedly reformulated system goals and asked for a much larger computer system than had been originally envisaged. In this chapter we focus on the first decision phase.

Strategies directed at achieving individual and group interests

In any situation where there is a conflict of interests, particular interests can only be achieved if an individual or group has sufficient power to impose a choice on others despite their opposition, or sufficient influence and persuasion to gain support for a particular point of view. In a democratic organisation persuasion is likely to be the most effective weapon in this search for personal support. In hierarchical structures such as industrial firms, the power which comes as a result of formal status, the possession of scarce resources and the ties of dependency between one group and another are likely to be one of the means used for this prosecution of personal interests.

In this context Rex (1961) has noted that 'if there is a conflict of ends the behaviour of actors towards one another may not be determined by shared norms but by the success each has in compelling the other to act in accordance with his interests'. If a number of individuals or groups have conflicting interests, are participating in the same decision process and are seeking, as one of their objectives, an increase in power and influence, this must lead to the creation of a very uncertain environment in which political strategies may play as large, or a larger part, than the rational choice mechanisms of the economists. This was the situation in our case study firm.

The strategies for influencing the attitudes of others which were most commonly adopted by each individual or group were, first, the use of *information* as a means for increasing the power provided by knowledge or as a means for persuading others that a conflict of interest did not exist. Second, the damaging of the *credibility* of those putting forward conflicting arguments through attempting to demonstrate that these arguments were not rational. Third, the securing of *support* through the creation of coalitions which enlarged the power base of the individual or group. Fourth, the *taking of action* at strategic moments when the opposition could not fight back.

Information as a power resource was recognised and manipulated by the subgroups within the Computer Department in their attempts to influence each other and top management. An early example was a discussion among the systems analysts on those political considerations likely to influence the decision at top management level which they must take into account if their equipment recommendations were to be accepted. Board approval and support were essential for success in the power game, therefore it was important for each group to identify those factors most likely to influence the Board in any recommendation that was made. During this discussion the Computer Manager entered the room and noted that the meeting was being taped. This implied that what was being said was of such importance that it needed to be recorded. As soon as he left the room the Systems Manager said, 'He saw the tape, we'd better get rid of it now. He'll be in here early tomorrow morning to play it back. We don't want him to get any pre-warning of our arguments.'

Although the systems group in this way attempted to keep strategic information from the Computer Manager, his position in the organisation hierarchy meant that he could easily foil the use of this information to influence top management. He did this by stopping the Systems and Programming Managers from meeting the Board. Previously all three had attended Board meetings together, but the Computer Manager now believed that the fact that the other managers' views on choice of computer hardware diverged from his own was increasing his personal insecurity in a high risk situation. He therefore ensured that in future only he had any direct contact with Board members.

The withholding of information was a device used by all groups as a means for reinforcing their own positions and preventing other groups gaining an advantage. This was true of the relationships between the programmers and the systems analysts. Uncertainty for the systems analysts was increased by the fact that they knew that the programmers could exert a great deal of influence on the choice of hardware, yet they had great difficulty in finding out how the programmers were assessing the situation. A systems analyst commented at this time: 'One of the big question marks for us is Dick's [the Programming Manager] position. We are never sure what his opinions are. We suspect he's heavily oriented to a Mitchell machine because of his experience with their software. BE (British Electric) software would mean a lot of change and trouble for him.'

At this time the programmers were quietly working on the machine and input solutions which would best suit their interests. The Programming Manager said: 'I re-did the volume figures for input and decided that teletype was what we wanted. We kept quiet about this for a time. Yesterday I went to see Jim [the Computer Manager] about it.'

The systems analysts had no inkling that the programmers were carrying out this kind of evaluation and the Systems Manager only heard about it the day before he went on his summer vacation. He favoured a totally different input solution but the programmers' timing of when he received the information meant that he was powerless to pit his arguments against theirs.

Another, and perhaps unconscious strategy, used by all individuals and groups, was the association of rationality with their own arguments and irrationality with the arguments of others. The Systems Manager was convinced that the Computer Manager was using emotional rather than carefully weighed arguments for the choice of his preferred computer. He said:

> Dick and I separately listed possible computers in order of preference. We both put Mitchell low down the list at around fourth or fifth place. [In fact the Programming Manager listed Mitchell as number 3.] BE submitted the cheapest estimate, around £100,000 less than the others. However, Mitchell is one of the top two choices. This is entirely because Jim [the Computer Manager] is emotionally involved with Mitchell.

The Computer Manager seemed equally sure that the analysts were not acting like rational economic decision makers. 'It's a fact that if some of our systems people had their way Mitchell wouldn't even be in the final selection. This is very largely a matter of prejudice. There's no doubt whatsoever of that.' He also viewed the programmers' choice of computer as being emotional.

> Dick's highly identified with ICM. He's a great friend of their local sales-people. He's still putting forward suggestions that their equipment should be considered. Dick's a pretty rational-logical person but his choice here is influenced by personal considerations, in particular his close relationship with ICM management.

The Computer Manager also perceived the Board as being influenced by emotional factors in their final choice of hardware: 'There were a number of factors which influenced the directors. They were not prepared to go along with ICM's proposals. This illustrates some of the anti-ICM feeling which exists in the Company. Any objection that can be raised gets a sort of murmuring consent immediately.'

As the decision process progressed and became more uncertain and stressful these accusations of irrationality become stronger. The senior systems analyst in charge of the project talked of resigning if the systems analysts' choice of hardware was not accepted.

> If Mitchell gets the order for the computer system and the reasons given by Jim are in any sense wilfully irrational – and they can only be irrational since BE have the best system – then I'm definitely leaving the Company. Somebody else can get on with implementing the system. I'll have had enough by then.

In this way the members of each group appeared to protect their own psychological security by refusing to accept that arguments which did not fit with their own had any logical basis. One reason for the prevalence of this emotional kind of thinking has been suggested in the last chapter. This is that an absence of precise and agreed criteria for evaluating choice decisions leads to many intangibles in the decision situation and an increase in uncertainty. This, allied with a strong belief that certain choices are better for personal interests than others, causes the decision process to have a strong emotional and political content and clear cut rational choice arguments do not exist to displace this.

Perhaps the most effective means for assisting self-interest is through altering the power structure of the decision process. This can be achieved by an individual or group securing the support of other groups for their interests; support which will be given either because these other groups believe that such backing will provide them with future rewards or because they perceive that there is some mutual identity of interest. In our case study firm the attempts of the systems group to protect and promote their interests were often frustrated by the development of a coalition between the Computer Manager and the Programming Manager. The Programming Manager seemed to recognise that his influence and the influence of his group depended on his maintaining good relations with the Head of his department; the Systems Manager used the riskier strategy of trying to make an impression at Board level even if this meant alienating his own manager.

Such a situation occurred when the Programming Manager came on a new form of input device which would fit well with the make of computer favoured by the Computer Manager. He told the Computer Manager about this and suggested it should be investigated even though this meant that he might have to abandon his own preference for an ICM machine. Although programmers rarely went to the systems section of the Computer Department one of them made a point of visiting the systems analysts and informing them of the new development. The analysts were thoroughly dejected after this news. As one of them put it, it looked as if all their work on input devices would be wasted 'because of some political manoeuvring between Dick and Jim and the manufacturer of the Mitchell computer'.

Soon after this alliance was formed, with the Programming Manager indicating that he would support the Computer Manager, the Systems Manager recognised that he was in a weak position and had lost influence. He reluctantly accepted the hardware and input choices of the Computer and Programming Managers, and for the first time in the decision process as it affected the Computer Department there appeared to be agreement on hardware. The Computer Manager began to hope that the differences within his group had been resolved and he attempted to reinforce the new identity of attitudes by holding a number of meetings with his managers at which they jointly discussed the best strategy with which to influence the directors. But although agreement was reached this stayed at a low level with the Systems Manager never fully supporting the opinions of the other two. He gave the impression of acquiescence without conviction. However, despite his Systems

Manager's lack of enthusiasm the Computer Manager wrote a report for the Board recommending the purchase of a Mitchell computer together with a teletype form of input device. The Programming Manager supported this course of action.

A unanimous recommendation to the Board therefore became possible because of an alteration in the balance of power in the Computer Department. All three participants recognised that this had come about as a result of political as well as technical factors. The Systems Manager said, 'Dick's still committed to ICM but he won't stand out on a limb and fight for his ideas when the head of the department is supporting another manufacturer. Dick is a person who has always liked a quiet life.' The Programming Manager later remarked, 'Bill [the Systems Manager] never had a chance with BE. When he realised this he pestered me to make a challenge for ICM. I was prepared to do this for my attitude was that if you are doing a job the size the directors were considering, then you had to look at ICM.'

This example shows that the furthering of personal interests through the decision process will not succeed solely through the skilful manipulation of information. The individual or group must also generate support for these propositions among key groups within the organisation. How they do this, who listens to them, how widely their ideas are diffused, are all critical factors in this generation of support. If a number of groups are involved in the decision-making process, all with different kinds of expertise, then a competitive struggle will develop. Thus the processing of demands and the generation of support are the principle components through which power within an organisation may be wielded. The final decision outcome will evolve from the processes of power mobilisation attempted by each party in support of its own demand. Much of the uncertainty of the decision process will be a consequence of the struggle for power and influence.

Effective political strategy not only involves the manipulation of information and the generation of support. It also requires an ability to prevent an opponent from taking protective or retaliatory action. A good example here was the disclosure by the Programming Manager of evidence that reinforced his group's interest on the day before the Systems Manager went on holiday. In this way the latter was rendered helpless and unable to protect his own position. For the remainder of the decision process the Systems Manager was always very nervous just before and during his vacation. Eventually he took to telephoning in while on vacation to check on what was going on. In this he was probably wise for during his absence the programmers used to visit the analysts and attempt to influence them to support the programmers' interests.

Once agreement had been reached within the Computer Department and a definite recommendation made to the Board of Directors, a marked change came over the Systems Manager and his analysts. Recognising a *fait accompli* they did a complete about turn and started putting forward complex accountancy arguments in favour of a Mitchell computer and teletype input. One analyst was heard to say to another, 'It's imperative Mitchell are seen to be thê best manufacturer for this job. We'll have to be real careful how we present the material. Nothing is to be put in the report which is in any way ambiguous, otherwise knowing the directors they might latch on to something and come out in favour of BE.

Yet only a few months previously a senior systems analyst had said that he would resign if Mitchell won the contract. The Systems Manager indicated that his days of aggressive challenging were over, or at least over for the present, and that he recognised the new power structure within the Computer Department. He said: 'I've decided to withdraw completely from the situation. I've advised the directors that a limited job with document readers would be the best approach. They've decided not to take my advice, which is their prerogative of course. I don't want the responsibility from now on. It's up to Jim to do all the talking and proposing.'

These are examples of the kind of strategies which were used to protect and enhance individual and group interest in our case study firm. Sometimes these strategies were effective, sometimes they were not, for a number of reasons which we can examine. It seemed that effective strategy was hindered by an absence of information which would enable an individual or group to assess accurately a political situation. There was frequently uncertainty about the validity of information that groups did have and there was a tendency to refuse to listen to information which did not fit with existing opinions and ideas.

There were two major information gaps which adversely affected the development of effective strategies by the groups within the Computer Department. Both concerned the Board of Directors. Until a comparatively late stage in the choice process the Computer Department staff were not certain that the Board really wanted the computer system they were investigating. They perceived the Board as ambivalent and would not have been surprised if the project had been summarily cancelled. Such a cancellation would have greatly affected the prestige and influence of the Computer Department and would have had a most damaging effect on its members. As well as this uncertainty about the future of the system the Computer Department were not in touch with the Board's strategic thinking and they made many attempts to find out Board attitudes and preferences. The fact that they were not very successful in doing this is illustrated by their surprise when the Board altered the system objectives. This absence of information concerning opinion at the top of the firm made it extremely difficult for the Computer Department to know what kind of information would carry most weight at Board level and to work out effective strategies for influencing the Board. A great deal of time and energy was wasted on fruitless search activities in which the information collected was quickly rejected by the Board.

Within the computer group political strategy was hindered because one group could not easily find out the intentions of other groups and individuals. Thus the systems analysts tried hard to discover the position of the programmers on hardware choice, while the programmers carefully concealed this until a time when they believed disclosure would assist their own group

interests. Similarly, both the systems analysts and the programmers needed
to know and understand the attitudes of the Computer Manager if they were to
persuade him that their hardware choices were superior to his. They certainly
did know the opinions of their manager as he never concealed these, but both
the Programming and Systems Managers tended to write these off as an emotional attachment to a particular computer manufacturer and for a long time
they refused to recognise that many of his arguments were based on perfectly
sound reasoning.

The Computer Manager stuck to his hardware preference throughout the
decision process but he was troubled from time to time because he was not
always certain of the validity of the information which he was receiving from
the computer manufacturer. At one point he expressed this uncertainty

> The Mitchell input proposals are not acceptable to us. They have pro-
> posed Class 33 teletype input machines which we wouldn't want to go
> along with as machines anyway, and we don't want on-line input, or any
> kind of punch operator controlled input. We have said to them 'Mitchell
> you have got to find another input solution.'

But, although both the Computer Manager and the Board of Directors did
have some doubts about the make of computer they were proposing to buy,
they kept these doubts at a low level and protected themselves from an increase in uncertainty by refusing to accept information which might add to
these doubts. For example, when the Computer Manager was worried about
the input proposals of Mitchell, British Electric managed to arrange a meeting
with him. The Systems Manager reported what happened.

> Ostensibly the meeting had been arranged to improve the contact between
> Jim and the BE higher management. Up until now 99 per cent of the
> contact between BE and this firm had been through my systems lads or
> I. . . . In fact, what happened was that Jim gave one of his speeches in
> which he more or less told BE his life story. He ended his speech with a
> burst of nervous energy half way through a sentence, told them that was all
> and threw them out. There was no chance for the BE people to participate
> in the meeting at all.

The Computer Manager's version of the meeting was less dramatic. 'I told
BE what the position was and pointed out that they were starting from way
back. They must not kid themselves that they were starting on equal terms
with Mitchell.'

The Board adopted more or less the same protective strategy as the Computer Manager when at a very late stage in the decision process they were
persuaded by ICM to agree to this computer manufacturer putting on a
special seminar to provide evidence of the advantages of ICM hardware. ICM
brought technical experts from all over Europe for this seminar. Unfortunately for them, only one director turned up and ICM were left talking to the
Computer Manager and his Programming and Systems Managers. In this way
an absence of information, a lack of certainty about the validity of information

and a refusal to listen to information that might increase uncertainty all affected the ability to size up situations and develop strategies that would successfully aid individual and group interests.

Strategies were directed at the manipulation of political situations, and individuals and groups were constantly having to make choices as to the approach most likely to further their interests. To an outsider they often appeared to make the wrong choice. This could be because of the information problems we have just discussed, but it could also occur because people had too restricted a view of the political situation; seeing only their small part of it and ignoring its wider ramifications, or being unable to identify the factors in the situation which were likely to exert the most influence on those groups which had the greatest power. Actions they might have liked to have taken were often rendered impossible because of their position in the organisational structure. We have seen that the Systems Manager needed to secure Board support for his proposals. Because these proposals did not fit those of his own manager the latter was able to prevent him having access to the directors. When the Systems Manager used the Computer Manager's holiday period to approach a senior director face-to-face, the Computer Manager on his return secured a promise from the director that this would not happen again.

The history of relationships as a factor in decision behaviour

Political strategy in our case study firm was greatly influenced by the perception an individual or group had of the interests and intentions of another individual or group. These perceptions were not shortlived responses to the present behaviour of the other parties but a consequence of the way relationships had developed over a long period of time. In order to understand the kind of social interaction that we observed during the decision-making process it is necessary to examine the history of relationships within the Computer Department. We would argue that several sets of past relationships did exert an influence on our case study innovative decision. First there was the Computer Manager's relationship with the manufacturer, Mitchell. Second, the history of the Computer Manager's relationship with the Systems Manager and, third, the relationship between the programming and systems analysts groups within the Computer Department.

The Computer Manager's relationship with Mitchell

Throughout the decision process the major argument used by the Computer Manager to support his demand for a Mitchell machine and to undermine the claims of British Electric and ICM was: 'This firm had a successful relationship with Mitchell. They have never let us down; why should I change?'

This stand of the Computer Manager was seen as illogical by the Systems and Programming Managers because of their memories of how the Mitchell relationship had developed. This perceived illogicality influenced their reactions to the Computer Manager's demand that another Mitchell computer should be bought, and reinforced their belief that other machines might be preferable.

Their perception of the situation was that the Computer Manager had developed strong personal ties with the directors of Mitchell and was reacting to these personal relationships rather than to the proved advantages of Mitchell computers. The researchers were told by one of the senior systems analysts:

> Mitchell more or less sold the machine to Jim. They had a lovely sales director called Henry de Ville. What a smooth character. I've never met anyone like him since. He came up here for a meeting with Jim, and I had to take him down to the 'Cat and Fiddle' for lunch with Jim and the chief accountant. Jim turned up $1\frac{3}{4}$ hours late. In this time de Ville didn't show the slightest sign of annoyance. His self control was unbelievable.

> He sold Mitchell to Jim. He told Jim, 'We'll fly a 350 over, here is a blank contract, have it on the conditions you want'. Relations between Jim and Mitchell were cemented over the next twelve months before de Ville cleared off back to France. Jim's really susceptible to the good salesman.

The Systems Manager said of the decision to buy the first Mitchell computer: 'It all boiled down to Jim's relationship with people at the top of Mitchell. He was on personal name contact with them.'

The Computer Manager was aware that he had this special relationship and it was recognised by Mitchell too, for one of their directors told a member of the research team that the Computer Manager was their best salesman. Nevertheless the Computer Manager argued that he was not being irrational. 'I do have a special relationship with Mitchell but it is not an irrational one so far as the company is concerned.' Certainly his emphasis on buying equipment that the firm already understood and found satisfactory appears very logical. Nevertheless, the fact that the other two managers were excluded from the cosy Mitchell relationship had a considerable effect on their behaviour during the decision-making process.

The Computer Manager's relationship with the Systems Manager

There were clearly personality differences between these two men. The first was a risk taker and an innovator, the second more conservative and cautious. However, much of their personal animosity can be traced to an incident which occurred some years ago when the Systems Manager applied for the O and M Manager's post and the Computer Manager opposed the appointment. The Systems Manager was given the job only after going over the Computer Manager's head to the directors. He believed that this subsequently affected the Computer Manager's attitude to him. 'I don't believe that Jim has ever forgiven me for getting the job. We mouth the right things. I had to go to the managing director; this announced to everybody that Jim's choice hadn't got the job.' The Computer Manager also believed that the Systems Manager still bore him a grudge for his refusal to support the latter's application for the job.

These kinds of personal difficulties may seem trivial in the context of a major innovating decision nevertheless our observations at the time were that they did have an impact on choice behaviour in the decision process and that, to some extent, the battle over the choice of hardware was a reflection of the personal battle between the Computer Manager and his Systems Manager.

Programmers versus systems analysts

The history of relationships between the programmers and systems analysts also influenced the behaviour of their managers in the choice situation. Over the years there had been a battle for status between the programmers and the systems analysts – these two groups were located in different rooms in the Computer Department and had little social contact. In the early days of the firm's use of computers the programmers were clearly the elite group. Computers at the end of the 1950s were extremely difficult to program and required specialists who had considerable knowledge of mathematics. At this stage the majority of the firm's programmers had good degrees in mathematics – some with first class honours. However, in industry generally, as computers have developed, there has been a change in the status position of programmers and systems analysts. Today commercial programming is not seen as a high level mathematical activity, but as something much simpler and more routine. Systems analysis, in contrast, is the new status area with its requirement that systems analysts shall be able to analyse complex organisational problems and to decide how computer based systems can be used to solve them.

In our case study firm, although the qualifications of the programming group have changed and none now has first class honours degrees in mathematics, programmers have held on to their status *vis-à-vis* the systems analyst group. This has been a result of the Programming Manager's leadership and the fact that he gives programming a broad definition and ensures that the systems analyst group does not erode any of the traditional functions of his department. In the innovative decision we are examining, the Programming Manager had a strong interest in maintaining the status and influence of his own group. The Systems Manager, in contrast, had an interest in using the decision process to extend the influence of his group and its share of company resources. The systems analysts recognised that their status was not as high as that of systems analysts in other firms and that many of the functions carried out by the rival programmers could be moved into their own area.

There was considerable evidence that the Systems Manager was interested in improving his own and his department's status and that he saw the proposed computer system as a means for doing this. If he could establish that his systems analysts were in charge of the project then he would be able to control it. He tried to insist that 'the current situation is that responsibility for any project lies with the systems analysts. There is no doubt about this, it's the stated policy of the department. The conception of the system and its implementation is the analyst's responsibility.' This was the first time that the Systems Manager had participated in any major computer decision, though the Programming Manager had been involved in two earlier choices of hardware. He therefore believed that if he and his group were to increase their power they had to do it through the present decision. They knew that the outcome of this decision would have repercussions in the Computer Department for many years to come.

Because of his own objectives the Systems Manager was very distrustful of the role played by the Programming Manager, in the choice process. He suspected that the latter might try to make decisions on hardware which would limit the analyst's systems design capabilities. The Programming Manager, for his part, played his cards very close to his chest and a problem for the systems analysts was that they were never sure what position he was going to take. This proved to be a very successful power maintenance strategy.

The systems analysts were therefore anxious to use the decision process to increase their own power position within the firm and the Computer Manager was very aware of this. He said: 'Their choice of manufacturer in the decision process wasn't just a matter of their technical orientation. Bill was actively concerned with putting forward a different installation, and one that was his alone.'

The Programming Manager too recognised the underlying motivation behind the Systems Manager's behaviour. He argued that the Systems Manager supported British Electric. 'Because he hasn't been connected with Mitchell in any way. He wouldn't like to admit they were good because he hasn't had anything to do with them. If we chose BE he could say "I decided alone". If Mitchell got the order the decision would have involved somebody else.'

Thus it would seem that a contributory factor to the conflict and uncertainty associated with our case study firm's decision processes was the attempt by the Systems Manager to increase the power and influence of his own group, countered by the efforts of the Computer Manager and the Programming Manager to maintain the existing task and power structure.

Reaction to technical and social uncertainty

The Computer Department of our case study firm had little experience of handling a project of the size they were confronted with. The Computer Manager acknowledged this in a memo to the Managing Director. 'The project we are about to undertake is of a different order of magnitude from our existing computer installations. We are moving into fields (online input, real time and multi-programming) of which we have little experience.'

The specialists within the Computer Department were well aware of the technical uncertainty of the environment within which they were operating. In addition they were faced with considerable social uncertainty. We have seen how the analysts were unsure of their relationships with the program-mers. One told us: 'There is still a tendency for the programmers to keep things to themselves. . . . It's difficult to know what they think.'

They were also unclear about Board attitudes to their work and this in-formation gap worsened when the Computer Manager excluded his two managers from meetings with the Board. A senior systems analyst commented: 'We're always complaining to Jim about being left in the dark. We never know what the directors think and I'm sure they have only a rough idea of what goes on in the Computer Department.' The Computer Manager too had his periods of uncertainty with the Board. At a late stage in the decision pro-cess a senior Board member said to one of the researchers. 'Of course we may decide not to go ahead with the new computer job despite the money we have spent on it.'

The researcher asked the Computer Manager if this remark was to be taken seriously? 'Yes it is. There is still a possibility that the firm will change its mind about the new system. If they do certain people are likely to resign.'

This uncertainty on the part of the Board may have partly reflected changes in its own structure that were taking place at that time. The old Board that had been with the Company since its formation was starting to break up through retirements and new directors were being recruited from outside.

But the major source of uncertainty for the computer specialists was the complex and dynamic technical environment surrounding the decision. Frischmuth and Allen (1968) have pinpointed some of the special difficulties of technical problem-solving. They refer to the fact that technical problems frequently have no terminal state; the solutions themselves are often dynamic. The Computer Manager was very aware of this.

> The great problem with this sort of decision is the uncertainty. Things are moving so fast you cannot be sure of anything. This means you've got to stop at some point and cut into the information, you've got to say 'This is the best possible decision we can make at this point' and then go ahead and implement this decision. If you're not careful there can be too many horizons in the field.

Frischmuth and Allen also argue that the goals which act as constraints on the decision process may change over time. We have referred to the fact that our case study Board changed the system goals for the new computer application some time after the technical search processes were underway. A further and critical source of uncertainty was that the criteria by which the solution was to be evaluated were dimly perceived and a matter for individual judgment. The Computer Manager wrote at the time:

> The configurations proposed by the manufacturers bear little resemblance to each other, and it is virtually impossible to compare them. . . . If one system had all the good features and another all the bad, at least it would be simple to compare the hardware, but examination of the various proposals suggests that the desirable and undesirable features appear to be randomly

distributed. What criteria is the client to employ in finding an optimum solution?

Given this uncertainty there is plenty of room for individuals, as we have shown, to reduce uncertainty by appealing to criteria which suit their personal interests. The Systems Manager's reply to a question of whether a decision to purchase a computer could ever be completely rational was: 'Not at all. Certain things can be evaluated quite rationally but there are also areas where no precise knowledge is possible. In these areas politics and emotion are bound to play a large part.'

External pressures on the decision process

In the last section we suggested that external groups with vested interests in the outcome of the decision process would attempt to intervene in it, and by doing so generate additional internal uncertainty and political reaction. The Computer Manager described to the researchers an interview he had had with a manufacturer's salesman shortly after three computer manufacturers had been shortlisted for further investigation.

'I had a visit from UDJ to find out why we had eliminated them from the tendering process. They expressed great amazement at the inclusion of BE and pointed out that it had no software and that ICM or some other manufacturer would have to be used.' The Mitchell salesman, commented on this kind of approach: 'The computer business is dirty. This sort of thing, this scandalmongering happens all the time, especially after companies have been put out of the running by a selection process like the one this firm has just had.' The Programming Manager was equally cynical: 'All the manufacturers tell blatant lies about their software capabilities. It doesn't annoy me. I just take it into account.'

A further tactic used by the computer manufacturers was to write to the Computer Department pointing out the enormity of the task it was setting itself and how the manufacturer alone was equipped to deal with such problems. The letter would invariably end with the announcement of some new piece of equipment. One manufacturer wrote: 'As we have discussed many times, this next state of computer development is not merely an extension in the size of computers you have already installed. It involves techniques and computer expertise with complexities of a different magnitude.'

Thus in our case study we see the mating process already described in an earlier chapter. Not only were the computer specialists searching for alternatives, alternatives were also looking for them. Because the Computer Manager had made it known that he was in favour of Mitchell, the other computer manufacturers concentrated their selling strategy on the Systems and Programming Managers. This added fresh impetus to the conflict and uncertainty in the Computer Department.

The Computer Manager expressed these feelings in his final memo to the Managing Director:

Because of the size of the order envisaged and the prestige attached to the installation, the Computer Department has been put under considerable pressure by rival claimants (computer manufacturers) extolling the virtue of their particular expertise. The difficulty of the situation is that few of these claims are subject to any scientific validation so that, paradoxically, one is left, in the selection of essentially scientific equipment, with large areas of belief. In this situation it is perhaps not surprising that agreement on selection of equipment does not even exist amongst informed opinion in the Management Services Department.

The Computer Manager therefore confirms the relationship we perceive between conflict and uncertainty, he also suggests an additional factor likely to promote conflict – the behaviour of computer manufacturers' salesmen.

Political activity of the kind we have been describing increased uncertainty and made the decision process more difficult to manage but it was not necessarily dysfunctional for the firm. Coser (1964) argues that power struggles which are a product of attempts to achieve personal objectives are often beneficial in their results, for a power struggle may lead to a new stability of relationships between the contending parties. He suggests that accommodation of divergent interests can only be reached after the contenders have assessed their separate strength in conflict. Eventually, 'the parties must agree upon rules and norms allowing them to assess their respective power position in the struggle. Their common interest leads them to accept rules which enhance their mutual dependence in the very pursuit of their antagonistic goals. Such arguments make their conflict, so to speak, "self-liquidating".'

This recognition of common interests was a feature of the decision process which we have been describing, though it led to a temporary rather than a permanent truce in the relationship between the Computer Manager and the Systems Manager, with the latter eventually leaving the firm. Because the strategies used by the Systems Manager to protect his interests alienated the Computer Manager, they rebounded back on the Systems Manager with adverse consequences. This is an example of how individuals, in pursuing what they perceive to be their own interests often succeed in securing an outcome which is in direct opposition to these personal interests. If decision-making behaviour is to be understood it is important for the observer to know how people perceive the situations in which they are operating, and to understand the extent to which the participants have accurate information on how their interests can best be achieved. In their attempts to prosecute various strategies, individuals continually commit errors because of misconceptions which arise through lack of information or through miscalculation. Individuals can also be manoeuvred into committing errors. Again, individual action may lead to unintended consequences. For example, somebody obviously striving for power may encourage others to consider their own interests more carefully and to take steps to maintain their own power position in the face of the new threat.

Conclusions on political behaviour in the case study firm

One of the most interesting aspects of the behaviour in our case study firm has been the manner in which individuals and groups developed a strong identification with a particular solution and refused to take account of evidence likely to demonstrate that there were viable alternative solutions. This behaviour pattern has been observed before. Devons (1950), discussing the formulation of aircraft programming in the Ministry of Aircraft Production in World War Two, tells us, 'there was a tendency for individual production officers to identify themselves with the firms for which they were responsible, and to regard as their main function the defence rather than the criticism of their actions'. Allen (1965) provides a psychological explanation for this behaviour. He quotes the work of Bruner and Postman (1949) on perceptual identifications and argues that the key explanatory variable is the cognitive system of the individual. A process of closure takes place whereby 'openness to additional cues is drastically reduced'. Dearborn and Simon (1958) also argue in this way concerning the departmental identification of executives: 'Presented with a complex stimulus, the subject perceives in it what he is ready to perceive. The more complex or ambiguous the stimulus, the more the perception is determined by what is already "in" the subject and the less by what is in the stimulus.'

Such explanations are interesting and perceptive but they tell us only *what* happens, rather than *why* it happens. We would argue that because organisational decisions are made by groups of people rather than individuals, and because they have complex processes associated with them, they will contain other stronger elements than those connected solely with the mental processes of the individual. Large-scale decisions are likely to threaten existing patterns of resource-sharing within the organisation. For those seeking to increase their departmental empires or personal power they provide opportunities for the acquisition of new resources – more influence, more responsibility, more prestige, more salary, for example. Although there is not a finite distribution of these resources within a company, nevertheless attempts by groups and individuals to increase privileges they already have are likely to be resisted by other participants in the decision process.

Questions of the allocation of resources have always been critical to the development of computer activities in our case study firm. With each computer system the Computer Department has grown in size and become more successful in imposing its own particular brand of technical rationality on a firm whose business success had been a result of good marketing rather than the use of sophisticated technology. In demanding this increased recognition of technical information as a business resource the computer specialists have critically affected the distribution of organisational resources such as departmental budgets and managerial career paths. Mumford and Banks (1967) have provided detailed examples of how computers can change expected career paths both to the advantage and disadvantage of different occupational groups.

Our case study Board of Directors, although prepared to extend the firm's use of computers, recognised that the increasing use of this technology in the business could act as a constraint on their own freedom of action. The Managing Director is said to have remarked to the Programming Manager just before he agreed to purchase the firm's third computer, 'I suppose it is too late to turn back, we can't do without them [computers] now.' Another director said: 'Intellectually I'm in favour of them, emotionally I'm all against them.' The Computer Manager remarked of this last statement: 'From his point of view why should he change? His empire is so neat – so tight. He knows every area of the business – how every last clerk works. Once we get in there with computers he won't know where he is.'

Thus any decision on how computers are to be used is likely to be influenced by interpretations made by individuals and groups on how their own position may be, or can be, changed as a result of the decision that is being taken and the choice amongst alternatives that is being made. Understandably there will be attempts to manipulate the decision-making situation so that personal interests can be served. To some extent this may be an unconscious as well as a conscious process, hence the tendency for selective perception and the mental rejection of information which does not fit with personal interests. These political issues appear to be important factors in how problems are perceived although clearly other variables also exert an influence. One is likely to be training, with the result that the technical man, trained to bring a technical view to bear on work problems, does not readily pick up relevant information of a non-technical kind.

We do not think that this kind of influence had a strong bearing on choice behaviour in our case study decision-making situation. Much more important were the attempts by the participants to maintain their own security and to increase their own influence and the influence of their groups. The Computer Manager was therefore attracted by a technical solution that he had confidence in and computer hardware that he knew would work. His own position *vis-à-vis* the directors of the company depended on the demonstrable success of a new computer system in a situation where there was still a mistrust of computer based solutions. The Systems Manager competed with the Computer Manager for influence with the directors but he could only achieve this by undermining the Computer Manager's position through producing a better alternative technical solution. In addition to this competition for power and influence the work his group had done on one input device meant that he had become very knowledgeable about this form of technical equipment. If this input solution were accepted then he would be the company expert, for no one else had this kind of information. The Programming Manager played a less active part in the power struggle in that he did not compete with the Computer Manager for the director's favour. But he had a strong interest in maintaining the status of his programming group in relation to that of the systems analysts. He also had an interest in protecting his group from the strains and difficulties of types of programming with which they were not familiar. His interest in the hardware choice was therefore in seeing that a machine was chosen which would do the job and yet would not be too difficult and demanding for his group to program.

But although all three men had different interests regarding the form of technology to be used, on the question whether the use of computers should be extended their interests became common. The more computers were used by the firm, the more new resources came into the computer specialists' environment and the more their power and influence increased. One of the lowest morale troughs that occurred in the Computer Department was when there was the suggestion from one of the directors that the whole project should be abandoned.

Summary

In this chapter we have discussed some of the political factors associated with the making of decisions and have provided an example of an innovating decision process concerning the choice of computer hardware. We have tried to show the social processes which led to political behaviour and the generation of conflict. The major research question asked of the data has been, Why did this degree of conflict occur?

Four theoretical assumptions have been put forward in explanation. The first, that organisational factors, such as the way tasks are allocated between groups, leads to a differentiation in norms, interests and objectives which will be reflected in any decision-making process. A large-scale innovative decision can be perceived by the participants as a means for promoting their own interests. For some individuals and groups this will be maintaining their own security through preserving the existing balance of power. For other groups a successful manipulation of the decision-making situation will be seen as a means for extending their own power position and for securing a larger share of those organisational resources which are attractive to them.

The second assumption is that existing relationships, which have developed over a period of time, predispose individuals and groups to behave in certain ways during the decision process. Thus earlier conflicts may be worked through again, using a new vehicle for their expression.

The third assumption is that conflict is a consequence of the uncertainty associated with most large-scale technical decisions. This uncertainty is generated by the number of variables that have to be evaluated in the decision-making process and by the essentially dynamic nature of this process with goals likely to be altered before a definite choice is made and with new technical possibilities appearing as the search process progresses.

The fourth assumption is that this uncertainty is increased by the behaviour of external groups, such as computer salesmen. In our case study the marketing strategies of the computer salesmen added fuel to the fire by continually presenting new technical options which could not be evaluated in any objective and scientific way.

In conclusion we would add that our experience in many firms where computer systems are being introduced is that the form of conflict we describe here is quite usual and a part of the stress and uncertainty of the innovator job. Usually researchers are not in a position to document this kind of conflict and after the event discussions with management on decision processes tend to focus on non-political issues.

PART FOUR
Uncertainty and the user

6

Innovator role definition and user uncertainty

In the previous two chapters we have outlined the kind of organisational uncertainty which the computer specialist has to face when contributing to large-scale innovative decisions concerning the use of computers. In the last chapter we concentrated particularly on the political factors associated with the internal decision processes and showed how these influenced the attitudes and behaviour of the computer specialists and, in doing this, increased further the uncertainty of the internal political environment. We now turn our attention to the kind of uncertainty with which the user department assimilating the new computer system has to cope. We argue that the manner in which the computer specialist defines his occupational role affects his attitudes and behaviour towards the user. If he adopts a narrow, highly structured definition of his role then this may reduce his personal uncertainty by restricting the number of factors for which he takes responsibility in the decision and implementation processes. At the same time a narrowly defined role is likely to increase the uncertainty of the future users of the system for they may believe that their personal interests are being overlooked and that the computer specialist is more interested in the technical efficiency of his system than he is in their job satisfaction.

The nature of the occupational role

In our society economic activities are clustered into occupations (Thompson, 1967). Organisations specify sets of activities which they require the individual in one of these occupations to perform while the occupation, in its turn, provides the individual with a sphere of activity in which he can use his skills and further his career (Mumford, 1972; Thompson, 1967). Most occupations have a different significance for the organisation and for the individual. The organisation arranges its division of labour in terms of historical custom and practice – that is on the basis of ideas which have become accepted over time on which activities should form part of the work responsibilities of

135

particular occupational groups. These ideas are often institutionalised and enforced through the influences of professional associations and trade unions. It also arranges them in terms of its technological requirements, its organisational structure and its value system on how work activities shall be allocated.

Thompson (1967) points out that certain occupations may be very tightly defined and have few discretionary areas:

> Jobs in the long linked technologies and in the protected portions of mediating technologies are highly standardised and repetitive, in part because such technologies can operate only when instrumental knowledge is highly developed, in part because such technologies are removed from environmental contingencies, and in part because organisational structure relates these jobs in relatively fixed patterns.

Some of these occupations may retain their original definition over long periods. Whyte (1961) describes the work of the glassmaker who makes fine glassware and shows how fabrication is still based on methods which are centuries old. Other occupations because they are affected by changes, particularly technological changes, in their environment, may be tightly defined but still subject to major change which causes one tight role specification to be replaced by another. The automation of certain parts of the car industry (for example, transfer lines) has had this effect.

Other occupations will change in different ways. Tightly specified jobs may become looser and contain larger areas of discretion. Cottrell (1962) gives the engine driver as an example here. When diesel and electric locomotives were introduced the engine driver was no longer required to have a 'strong back and a weak mind' and a knowledge of automachinery became a critical job skill. In recent years some computer systems have had the opposite effect to this, taking away discretion from the clerk and giving him a much more tightly specified set of duties than previously.

Boundary-spanning occupations such as those carried out by most specialist groups vary considerably in the amount of discretion they contain and in how this alters as the occupation develops. Thompson suggests that the way occupations of this kind are redefined over time depends very much on the nature of the environment at the boundary of each occupational role. If this is homogeneous – that is, does not contain many different influences – and also relatively stable, then these kinds of jobs can become standardised and a set of clearly defined skills attributed to them. At present commercial programming appears to be moving in this direction as parts of the programming package are built into the machine. If, in contrast, the occupation exists in an environment where it has many different influences impinging on it, then it will have to be more adaptive. It may be subject to pressure from labour and product markets, from diverse demands within the firm and from complex technology and if this is the case it will have to alter its activities in order to respond to these differing stimuli. If, in addition, the environment is also dynamic, for example, the technology the occupation interacts with is developing or the other occupational groups which it services are changing the nature of their demands, people holding this kind of occupational role will be required to exercise a great deal of discretion and to be constantly redefining their role and rethinking the set of activities incorporated in it.

In order to understand how occupational roles have originated and developed it is important to identify those influences in their environment which have led to a role being defined in a certain way and to particular sets of activities being associated with it. It is also important to examine historically how these factors have altered over time, causing the role to be redefined both by the organisations in which the occupational role exists and by the individuals who occupy the role. Sociologists have sometimes tended to forget historical influences when explaining organisational behaviour and to ignore the opinion of Homans (1967) who says, 'The past affects the very way the future comes into being', and of Nadel (1957) who argues that 'we must not forget that in "real" situations roles-relationships have a significant pre-history; they are not created *ex-novo* in the context in which they are observed to operate'.

We believe that an understanding of occupational role structure and the historical development of an occupational role is essential to an appreciation of why individuals and groups behave as they do. Our reasons for approaching the behaviour of the computer specialists in our firms from this position is set out in the next paragraph. But before embarking on a discussion of the activities of the computer specialist it is necessary to state what we mean by the phrase 'occupational role'. Here we use a definition which follows Becker and Carper (1956), Strauss *et al.* (1964), Reid (1968) and Bucher (1970). This sees occupational role and occupational identity as involving the following components:

1 a definition of the field with which the occupation is identified – its boundaries, its major body of knowledge and associated methods;
2 a mission which the field serves – the value system which justifies and sustains the occupation;
3 the activities which are proper to the field;
4 the relationships that should obtain both between members of the field, and with persons in other fields.

The many studies of computer introduction which we have now made suggest that three factors have a considerable influence on the initial attitudes of employees to this form of technical change and on the degree of uncertainty they feel about it. The first of these is the judgment which employees make on the likely effects of the change. This judgment is influenced by impressions derived from sources external to the firm such as books, newspaper articles and television, and by internal sources such as the observed effects of computer systems already operating in the company; but it is also affected by interpretations concerning the role of the groups most involved in designing and introducing the change, namely the computer specialists. Our second major factor influencing user department attitudes is therefore the way staff there perceive the occupational role of the computer specialists. If this is seen as generally benevolent and sympathetic to user department needs, then, although the introduction of computer specialists may not be actively welcomed, their visit will not be viewed with hostility. If, in contrast, their role is interpreted as opposed to the real interests of the user department, however these are defined, and the computer specialists are viewed as hard technical men whose sole interest is in rationalisation through EDP irrespective of the effect of their systems on the non-technical needs of user staff, there will either be outright opposition to their activities or, if the user department lacks the power to resist, anxiety, uncertainty and covert resistance.

The third factor indirectly influencing the attitudes of the user department is the perceptions which computer specialists have of their own role, and the way they define their activities and responsibilities and translate these definitions into strategies for action. The reasons why occupational groups both define their role in certain terms and have their role defined by others in either the same or different terms is an interesting and relatively unexplored area. Interviews and discussions with computer technologists suggest that their role perceptions at any one moment in time are influenced by the factors set out in Fig. 6.1.

Starting at the righthand side of the figure, company expectations and demands refer to the influence on computer personnel of the way the employing firm defines the role it wishes its computer specialists to play. This definition is governed by senior management decisions on how computer

Fig. 6.1 Factors influencing occupational role definition

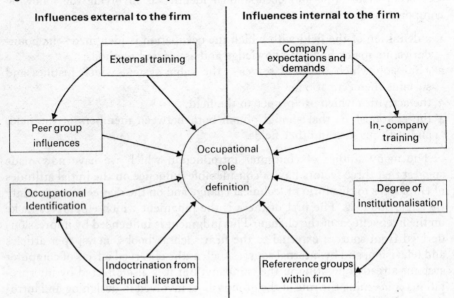

technology is to be used and developed, and these policy formulations are, in turn, a response to external pressures from product and labour markets with computer technology seen as a means for enabling the firm to respond better to these pressures (Mumford, 1972). Many managements have worked out explicitly how their firms will use computers in both the short and medium term and this clear definition is shown in detailed job specifications for programmers and systems analysts, in precise job advertisements, and in carefully designed in-company training programmes. (Few, if any firms, will be able to forecast their use of computers in the long term. There are too many unknowns in computer technology.)

Other firms, perhaps more newly associated with computer technology, are less sure of their requirements and because of this they may hire computer specialists but leave them largely to define their own roles. When this is the case the influence of the external factors in the boxes above may be stronger than that of the internal factors. The computer man may try to become what he wants to become, or what he thinks it is desirable career wise for him to become rather than what his firm needs him to become if its objectives are to be achieved.

A computer specialist's reference group(s) is that group or groups whose approval or support is important to him and whose expectations will therefore exert a strong influence on his behaviour. The kind of influence which he responds to is affected by our third factor, degree of institutionalisation. By this is meant the length of time an individual has spent with a particular employer and the extent to which he has become identified with overall company interests and acceptive of company value systems. We have found that where computer specialists are young and recruited straight into computer departments without other job experience they take as their principal reference group their colleagues in the computer department, and their behaviour is shaped more by the values of these technical specialists than by the values of employees located elsewhere in the firm who become users of their systems.

In contrast, where computer specialists are long service employees recruited into the computer department from other parts of the firm they are already indoctrinated with company values, and their reference groups are as likely to be the staff of the user departments where they are introducing new computer systems as their own colleagues in the data processing department. We have found that the choice of reference groups which the computer man makes, together with the extent to which he is institutionalised into the value system of his firm through long service, has a considerable impact on his approach to systems design and on the kinds of strategies which he uses for getting new computer systems implemented.

The computer specialist not only has his occupational role moulded by internal influences within his company, he is also given expectations of his task responsibilities and the parameters of his job by a number of influences external to his firm. These are shown on the lefthand side of Fig. 6.1. One of the most important of these is the kind of vocational training he receives in educational institutes and the stress this gives to different aspects of his work. Until now, the computer specialist, like the engineer, has been given a training which focuses almost exclusively on the technical elements in his job and ignores the fact that all technical change involves altering attitudes, social structures and work activities, and therefore has a large non-technical design element. But whereas most engineering courses now include instruction in the social sciences, this is still not true of the majority of computer courses. Despite the length of time the knowledge of human needs and behaviour provided by sociology and psychology has been available it is still possible to attend meetings of computer experts at which the view is authoritatively expressed that there is no systematic body of knowledge in the human area which can be imparted to computer systems designers. While those who design training courses believe this and incorporate in their syllabuses only the technical aspects of the system design job, we may expect that computer specialists will be strongly influenced to define their functions solely in technical terms.

This emphasis on systems design as a purely technical activity is reinforced by external peer group influences such as attendance at meetings of the British Computer Society. Although social issues are raised at these they tend to be related to specific problems such as computers and 'privacy', rather than to discussions of how computer systems can be designed so as to improve the human quality of life. We would argue that the systems designer should now be enriching work rather than routinising it, increasing the control of the individual over his work activities rather than increasing his subjection to tight external controls, providing more rather than less responsibility, increasing rather than reducing job satisfaction. As yet, these things are rarely seen as part of the design responsibility of the computer specialist. There is nothing in his training which gives him a set of design principles for handling these human areas, and little in his contacts with his fellow computer professionals which makes him think that this knowledge should be part of his design expertise.

Technical influences are reinforced by computer literature. This too concentrates its attention on the technical skills associated with computer work and pays little attention to other kinds of skills. Where problems of human beings are discussed there tends to be an emphasis on a marketing approach. Systems must be 'sold' to users; users must be 'persuaded' to accept them. The human part of any system is not seen as a design variable which requires skilled knowledge if it is to be handled successfully.

The last factor influencing the way computer specialists define their role is the extent to which they identify with electronic data processing as an occupation. Do they intend to make it a lifetime career or is it a set of skills which they wish to acquire before moving on to another kind of management job? The career computer specialist may define his role in rather different terms from the individual who is anxious to move on to a general management position.

The computer staff in our four firms

The EDP departments of our four firms differed considerably in size. This was not due to their functions being markedly different but to the fact that our manufacturing firm, Falcon Ltd, and the government service department, AEK, included data preparation staff such as punch operators in their departmental complements, whereas the distribution firm, Poultons, and our other manufacturing firm, Grant and Co., did not. Because this chapter is concerned with the way computer specialists define their roles when designing and implementing new EDP systems we shall be concerned only with the programmer/systems analyst group and a great deal of our attention will be directed at the occupational role perceptions of systems analysts as this is the group which has the most direct contact with the user departments.

At the time of our study Poultons employed thirty-four computer specialists – seventeen programmers and seventeen systems analysts; Falcon Ltd, also employed seventeen programmers and seventeen systems analysts. AEK, which was some distance away from the implementation of its computer system, employed seven systems analysts and twenty programmers. Grant and Co. had a large management services department at group level, so the local complement consisted of only two systems analysts. All the systems analysts we interviewed told us that they had responsibility for analysing existing methods of work in the company and for recommending new and improved methods which would generally, but not exclusively, involve the use of computers. This meant that our systems analysts had to use observational and analytical skills to understand the workings of the existing manual systems. Then they had to weigh up the deficiencies of these manual systems and decide how they could be improved, using the potential of the computer to assist this process. By improvement the systems analyst generally meant making the workflow through a department faster, more accurate, and more responsive to user management demands. He did not define improvement in the sense of increasing job satisfaction or producing better working conditions for user department staff, although on occasion these could be the products of his system.

Once a computer based solution to a business problem had been formulated the systems analyst had to design the new procedures which formed part of this solution and, depending on where the boundary between programming and systems analysis was drawn, he might have to define the programming requirements. He had to ensure that staff in the user department were trained to operate the new system and he, together with the user manager, had to ensure that the system was successfully implemented.

We have already indicated that the systems analysts' role was defined both by the systems analysts themselves and by those other groups who interacted with them, in terms of a set of technical design activities. Human design activities which might legitimately have been incorporated into this role definition were rarely to be found there. Evidence of this technical orientation was obtained through asking the systems analysts in our four firms to define systems analysis for us and to describe what the job involved. Typical answers were:

> It is the solving of problems, basically with a computer solution in view. It involves detailing the problem, proposing a solution and specifying how jobs will be done with the computer, both by machine and manually.

> To produce the required output from the given input while minimising the sum of computer time costs, related labour costs and machine costs; yet allowing flexibility and practicality and ensuring accuracy.

Not one answer made reference to human factors in this design process, other than to the need to reduce labour costs and ensure accuracy.

Internal factors which affected role definition

The influence on systems analysts' role perceptions of company demands and expectations

In all four firms there had been a traditional clerical technology in those work areas where we were studying the introduction of computer systems. Computers had been introduced with the objective of making clerical procedures more efficient and less expensive, and our firms hoped to achieve savings in the number of clerical staff employed, although this was nowhere the principal reason for investment in computers. The use of computers was seen as a means for increasing company efficiency and profitability, or, in the case of the government department, of increasing efficiency and reducing costs. It was in these terms that the firms viewed the role of their systems analysts. Although each organisation wished the change to be handled in as humane a manner as possible with staff adequately informed on what was to happen, no positive social goals, in the sense of attempts to improve work interest or increase job satisfaction through a humanistic design approach were included in system design objectives, though many systems analysts, particularly within AEK, did hope that their efforts would lead to a reduction in routine work and therefore to an increase in job satisfaction. This is not to suggest that any of these firms were disinterested in the welfare of their employees. On the contrary all had similar human relations philosophies. With the exception of Grant and Co. which was a small firm, all were large and paternalistic with reputations for treating their staff fairly and with consideration. These firms merely represented the traditional view, which still holds good in Britain, though it is now changing in the Scandinavian countries, that the interests of employees are best looked after through paying fair wages and providing good working conditions and fringe benefits. Employees were expected to conform to the requirements of technology and, if a new work system was introduced to take advantage of the benefits of computer technology, then they had to adjust to the demands of this system.

The firms' expectations of their systems analysts were therefore that they would exploit the potential of computer technology to increase efficiency and facilitate the achievement of commercial objectives (client service objectives in the case of AEK), and that they could communicate to employees what they were doing and try to win approval for this. There was no requirement for any of the systems to be designed so as to secure direct positive benefits for the employees who would operate it.

This technological and commercial value system was made apparent when we asked the managers of computer departments in each firm what factors would weigh most heavily with them when they were considering a new use for computers. They told us that these would be,

- the departmental savings (staff or cash) likely to result from the new idea;
- the return on investment the company might gain from the proposed innovation;
- the technical difficulties of getting the new idea into operation.

Thus all our firms operated from a similar value position in relation to the use of technology. That is, a wish to exploit the potential of the technology and a belief that employees would accept any new tasks required of them. They were in no sense to be blamed for taking this position. All their experience demonstrated that staff would conform to the demands of a new work system; the trade unions to which their employees belonged still saw earnings as their principal bargaining area and had not enlarged this to cover other aspects of work such as job interest or job satisfaction; the Civil Service Unions were an exception here. The firms' prime requirement of their systems analysts was technical competence. In turn, the systems analysts they recruited, for a variety of reasons discussed below, also defined their responsibilities in terms of designing technically efficient computer systems.

It can be suggested that all these firms had a confused personnel ethic and reflected the conflict in society between 'human' and 'technological' values. They wished to be good employers but were strongly influenced by the desire to achieve technical gains. Ellul (1965) has suggested that old ideas of scientific management are still exerting a sinister influence on modern society; here the enemy is not the machine or any particular technology, it is 'technique', or the drive to rationalise every human activity; the search for the 'one best way' of Taylorism. He maintains that technique tends to take over all man's activities, to transform everything it touches into a machine. Although our firms required their systems analysts to show a concern for human relationships, this was restricted to ways of gaining acceptance for the proposed change There was no requirement for systems analysts to have the skills necessary to design systems which would enhance the quality of life for workers as well as improving company efficiency.

The reference groups of the systems analysts within the firms

An important influence on how an individual defines his own role, as opposed to how others define it, is the nature of those groups within the company which exert an influence on his attitudes and behaviour. The systems analyst, in contrast to the programmer, interacts more with client groups using computer systems than he does with colleagues in his own department. Figure 6.2 shows that this was the case in our firms.

Fig. 6.2 Groups with which data-processing staff said they had the most contact

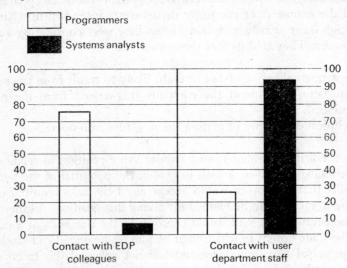

Frequently these groups have very different value systems. Whereas most computer specialists are wedded to the technical ethic we have already described, many user departments have other objectives which they see as equally important. These are likely to include creating a pleasant working environment, establishing good staff relations, avoiding stress and anxiety, as well as operating economically and efficiently.

We found that our systems analysts did recognise the divergence in goals between their own and user departments, though they did not always understand the reasons behind this divergence and tended to attribute it to the inability of user department staff to perceive the benefits of computer systems. Seventy-six per cent of systems analysts believed that a user department's attitudes towards computerisation differed from those of the computer department. We were told:

They have vested interests and often an aversion to change.

Yes, our aims are different. Most users want to do a day to day job. The pressure on them is such that they cannot look for better ways to do it.

Yes, there is a lack of appreciation of the potential of computers.

If the computer man has only one principal reference group, his own department, he is likely to pursue his technical goals without paying too much attention to the goals of the user department. If, in contrast, he is closely identified with the needs and interests of his user clients he will define his role more broadly and see as part of his functions the necessity for not making user goals more difficult to achieve.

We asked our systems analysts if they ever experienced a conflict of loyalties between the objectives of their own department and the needs of the user department. We found that 41 per cent did experience such a conflict and that 18 per cent found this kind of conflict to be unpleasant. They had different methods for resolving it Some would side with the user department's interests, others would support the approach of their own department. Figure 6.3 sets out our definition of the effect on systems analysts' behaviour of having computer specialist colleagues as the only reference group, compared with taking both computer and user departments as equally important reference groups.

Fig. 6.3 Systems analysts reference groups

Systems analysts	User and EDP goals conflict	User and EDP goals do not conflict
User dept. is reference group High congruence with user goals	Consultation, participation Attempts to get system modified to take account of user goals ∴ reduction in uncertainty Stress Role ambiguity of systems analyst	Consultation, participation User involvement in systems design ∴ low uncertainty Harmony Role clarity of systems analyst (human relations content recognised)
Computer dept. is reference group Low congruence with user goals	Marketing approach-selling persuasion System designed to meet technical goals ∴ high uncertainty Authoritarianism Role conflict	Demonstrate system will be of value System designed to meet technical goals ∴ uncertainty may be high or low Proof Role clarity of systems analyst (technical elements predominate)

If the systems analyst sees the user department as an important reference group and is aware of, and identified with, a broad range of its interests and objectives, then he is likely to take account of these when designing and implementing his computer system. If there is a conflict of goals and the achievement of technical goals will threaten some of the department's existing social goals, then he will deal sympathetically with this problem and attempt, wherever possible, to modify his system to meet user department needs. Nevertheless this conflict of interests may mean that his role is not entirely clear to the user department. They will not be certain whether he is for them or against them.

If the systems analyst takes the user department as his reference group and there is no conflict between the goals of the computer group and those of the user department then there is likely to be a harmonious relationship and little uncertainty about the human outcomes of the innovation. User staff are likely to be actively involved in the specification and design of their own systems. They will clearly understand the role of the systems analyst, will have worked out with him the boundaries between his responsibilities and theirs, and will perceive his role as containing a large human relations element.

But if, in contrast, the systems analyst takes his own computer colleagues as his principal reference group and has little sympathy for, or understanding of, user department needs and objectives, then he will define his role in very different terms. If there is a conflict between his technical goals and the human relations goals of the user department he is likely to cover this up through marketing strategies in which he attempts to sell the virtues of his computer system, while concealing its threatening aspects. If this marketing approach does not work then he may seek to use more authoritarian methods, getting, if he can, top management sanction to ensure that his system goes in essentially as he has designed it. This is a high uncertainty situation for the user department. If the systems analyst finds that he is in the fortunate situation where, although his system has not been consciously designed to meet the user's non-technical needs, it does not threaten these, then his objective will be to demonstrate to the user department its efficiency and advantages. His role will still be defined in technical terms but, because the values and interests of the user department are not being threatened in any way, this role definition may be acceptable and will not generate a great deal of uncertainty for the user.

Although an individual's personality and temperament must influence his awareness of the needs of others, and systems analysts are no different from the rest of us in this respect, our evidence suggests that the young computer man who moves straight into a data-processing department without working anywhere else in the firm, tends to take his own department firmly as his reference group and has difficulty in perceiving or understanding the non-technical goals of the user department. In contrast, systems analysts who, as in Falcon Ltd, and AEK, have worked for many years in jobs which have nothing to do with computers and, as in AEK, may have been located in departments which they later help to computerise appear to have a much better understanding of the social goals which are important to user departments, and are less likely to be hardline technical men.

At present it seems that personality and previous experience outside EDP are the principal factors which lead systems analysts to broaden their occupational role definitions and to perceive the non-technical goals of the user as being their concern. We would hope that in the future new forms of training will also assist this broader role definition.

Institutionalisation as a factor in occupational role definition

There were differences in the way in which our firms recruited staff to their computer departments. Whereas Falcon Ltd and AEK recruited internally, Poultons brought in the majority of their staff from outside, and Grant and Co. obtained staff from Group Headquarters. This policy affected the age structure of the computer departments and the length of time individuals in these departments had spent with their firms. In Poultons the computer staff were young; 70 per cent of them had less than four years' service with the company, and all this service was in the computer department. Only 11 per cent of Falcon Ltd, and 4 per cent of AEK's computer staff had less than four years' service and both these organisations had some EDP staff who had worked for their employers for twenty years. If all the firms are put together the average length of time in computer work was $3\frac{1}{2}$ years, but, whereas 47 per cent of the Poulton systems analysts had worked only in the computer department of that firm, no Falcon Ltd or AEK systems analysts had worked solely in data-processing. All had been in other kinds of jobs in their organisations before becoming involved with the computer work. Many of these long-service employees had spent all their working lives with their present employers. We found that 53 per cent of Falcon Ltd, and 71 per cent of AEK systems analysts had had no other employer.

There was therefore a considerable difference between the job histories of systems analysts working in Poultons and those working in Falcon Ltd and AEK. Because the systems analysts of Falcon Ltd and AEK were more institutionalised (they had longer company service) than those of Poultons we should find differences in the way they defined and carried out their job responsibilities. We would expect them to be more imbued with the overall value systems of their companies and more aware of the human relations needs of user departments – after all they had worked in these departments themselves. They should therefore see their occupational role as having a larger human relations content than the systems analysts of Poultons who had a shorter length of service and had never worked outside the computer department. This proved to be the case. When we asked systems analysts to define their role in terms of its technical and human relations content, they replied as shown in Table 6.1. Grant and Company systems analysts are not included because they were only two in number.

Table 6.1 Systems analysts' role: technical and human relations content

	ROLE HAS TECHNICAL CONTENT ONLY		ROLE HAS TECHNICAL AND HUMAN RELATIONS CONTENT	
	n	%	n	%
Poultons	9	53	8	47
Falcon Ltd	–	–	17	100
AEK	1	14	6	86

How these different role perceptions affected work behaviour is discussed
below, when we examine the systems analysts' answers to questions on how
they got their computer systems accepted and implemented.

Training

A major influence on the way the members of occupational groups define
their work activities is the kind of training they undergo to gain entry to the
occupational group. Typically, agencies which provide such training are
effective not only at establishing minimum standards but also at controlling
entry into training and thereby the supply of those so trained (Thompson,
1967). Often training appears to fix an individual in a certain perception of
his occupational role and it becomes extremely difficult for him to alter this
perspective even though a changing work situation means that this role is
wrongly defined for new organisational needs. The very fact that training
programmes are set up implies that there is some consensus, at least amongst
the group responsible for designing the syllabus, on what the content of the
role should be. If the occupation has a long historical tradition like, for
example, medicine or law, then community expectations which have developed
gradually over a long period of time will influence decisions on what should
be learnt. However, there is often a timelag between community expectations
and training modifications. For example, most people would see a need for
doctors and dentists to have a sound knowledge of human psychology if they
are to meet the mental as well as the physical needs of their patients. Yet it is
only comparatively recently that psychology has been included in medical
courses and there are still many dental courses on which it is not taught at all.
Eventually, however, most training programmes respond to community
expectations of what should be the responsibilities of people holding particular occupational roles and to the perceptions of the people in these roles on
where their responsibilities should begin and end.

The two principle occupational roles associated with computer technology
– programming and systems analysis – have always been difficult to define and
have only recently had professional qualifications associated with them.
Agreed job definitions for systems analysis are notably hard to find. An article
in the *Computer Weekly* (Maunder and Styles, 1969) has pointed out that the
development of training programmes has been handicapped by an almost
total lack of agreement on the functions of a systems analyst. It suggests that
at times it has seemed that the only common ground was that he appeared
somewhere between the user of a system and the computer. Within that area
the boundaries of his responsibilities were almost infinitely variable.

Today, computer specialists can obtain professional qualifications by sitting
the examinations of the British Computer Society or by taking a university
course in computer science. At the time we carried out our survey the British
Computer Society examination scheme was only being formulated and although a few computer science degree courses were in existence these were

not seen as the means by which a commercial systems analyst should qualify himself. The practice was to move into the job without any formal qualifications in computer science and to concentrate on obtaining theoretical knowledge through attending short courses. Any indoctrination on the nature of the systems analyst's occupational role obtained through training was not therefore in terms of a clearly defined corporate body of knowledge associated with a set of professional values but much more *ad hoc* bits and pieces of information obtained when time permitted, in an attempt to fill up perceived gaps in computer knowledge and to keep up with a rapidly developing new technical specialism.

Twenty-seven per cent of the systems analysts in our four firms had university qualifications, though the distribution of degrees was not even between the firms. Falcon Ltd employed the highest percentage of graduates and 47 per cent of its systems analysts had degrees. Poultons came next with 35 per cent of systems analysts holding degree qualifications, whereas AEK because of its policy of recruiting staff after school 'A' levels, had no systems analysts with a degree. Poultons tended to recruit graduates with science rather than arts degrees, whereas a majority of Falcon Ltd systems analysts had arts degrees. It will be remembered that Falcon Ltd had a policy of moving people into systems analysis from other jobs and did not recruit direct from university. We found that training for data-processing was everywhere organised or encouraged by the employer. In all four firms virtually no one had taken a data-processing course of any kind before joining the company. Thus systems analysts' perceptions of their occupational role responsibilities were derived from training obtained during employment with their present firm.

Once employed as a systems analyst, however, courses were embarked on with much enthusiasm. Table 6.2 shows the percentage who had been on courses. The average number of courses taken was three but the range was from one to ten. Here again we have omitted Grant and Co. from the table.

Table 6.2 Percentage of systems analysts who had taken courses

	% OF SYSTEMS ANALYSTS WHO HAD TAKEN COURSES PROVIDED BY THEIR EMPLOYER	% OF SYSTEMS ANALYSTS WHO HAD TAKEN COURSES NOT PROVIDED BY THEIR EMPLOYER
Poultons	71	24
Falcon Ltd	76	36
AEK	86	86

These courses were in different aspects of computer technology perceived by either the employer or employee as relevant to the work of the systems analyst. Many of the systems analysts we spoke to had every intention of taking more courses in the future.

Table 6.3 Systems analysts – future course attendance (percentages)

	INTENDS TO TAKE FURTHER COURSES	DOES NOT INTEND TO TAKE FURTHER COURSES	NOT SURE
Poultons	47	18	35
Falcon Ltd	59	12	29
AEK	43	43	14

This pattern of course attendance makes it difficult to determine the influence training had on the way our systems analysts defined their occupational role. The majority of these courses were short and on data-processing techniques. It seems likely that this type of course reinforced the systems analyst's perception of his role as a primarily technical one.

External factors which may affect the role definition of computer technologists

Group stereotypes which exert an influence on occupational role definition

Most occupational groups compare themselves with other occupational groups whose activities they see as similar to their own. Frequently these comparisons are made to assist bargaining with employers over wage and salary levels. For example, in the past, university lecturers have used senior civil servants as a reference group against which to compare the equity of their salary structure. But occupational reference groups may be used for purposes other than bargaining. They may be taken as models for behaviour because their activities and values are respected and looked up to. In addition there will be other groups in society who exert an influence on the behaviour of members of an occupation. These groups are not usually admired or taken as models, but, because they have become interested in the behaviour of the occupational group in question, they adopt a critical stance concerning what the members are doing and how they are doing it; and they can act as a stimulus for behaviour change.

Groups which computer specialists are likely to take as models will be other scientific and quasiscientific groups in industry, such as operational research and research and development. Many of these will be staffed by professional scientists who have been indoctrinated in scientific values and norms of behaviour and who may have a powerful sense of community. Computer specialists do not yet have the characteristics of these professional scientists, though they are being moved in this direction by the British

Computer Society with its professional qualifications and code of professional ethics.

This is not to suggest that the computer specialist in any sense models himself on the academic scientific community. The latter role is usually seen as based on a strong set of values incorporating ideas on the importance of 'objective' truth, the free availability to all scientists of scientific findings, a concern for objectivity in scientific activity and an organised scepticism (Merton, 1949). Most scientists working in industry accept that the objectives of commercial organisations may require a dilution of some of these values, especially that of the free availability of knowledge. Nevertheless the industrial scientist still seeks after truth, although this may be a focused problem-solving truth rather than a generalised theoretical truth. He still attempts to take an objective view and weigh up evidence carefully and accurately. But he may have to accept that his solution will be taken over by others and modified to suit the needs of his company rather than the value system of his academic discipline. The computer specialist appears to have adopted a set of values very similar to these. He attempts to analyse problems logically, analytically and truthfully; he has a strong feeling for the importance of technology, especially his own technology and the advantages its use can secure for his firm. Whenever possible he likes to be tackling difficult problems and to be working close to the frontiers of his discipline. He hopes to be able to implement his solutions essentially as he formulates them, though the necessity for taking account of the interests of other groups, especially those of departments which use his computer systems, may make compromise inevitable. The value system of the computer specialist therefore appears very similar to that of the industrial scientist, even though it is not as yet compulsory for the computer man to undertake a rigorous scientific training before acceptance into his occupational group.

It is not easy to establish whether the computer specialist has adopted the norms of the industrial scientist because he accepts these as a desirable model or because the demands of his innovative occupation lead logically to a set of values of this kind. Historical evidence suggests that both influences may play their part.

In its beginnings (the early 1950s), computer technology was dominated by programmers and, because computers at that time were very complex pieces of machinery, their programming required considerable mathematical ability. The first programmers therefore tended to be well qualified mathematicians. Because they were a new occupational group introducing a technology which was new to industry and often greeted with suspicion and hostility, they became tightly integrated and mutually supportive, and many made this clannishness apparent by adopting a set of behaviour norms that set them apart from other groups in the firm. In one of our firms this took the form of wearing unconventional dress and working unconventional hours; both these activities form part of the stereotype of the industrial scientist. As computer technology advanced, programming became simpler and a new group appeared on the occupational scene, the systems analysts. This was an entirely new job for which no qualifications existed and systems analysts needed to acquire credibility quickly if they were to get their role accepted.

Programming now moved into its next stage of development, becoming technically easier at the commercial level while the market demand for programmers increased rapidly. These two factors reduced the standards of experience and training which firms required and could expect to find, given the state of the labour market. This meant that there was now an occupational group in existence which, because of its history, still had norms of behaviour similar to those attributed to the industrial scientist but which was not required to undertake the kind of long-term formal training programme which would provide it with a set of professional skills and values. The new systems analyst group gradually overtook the programmers in status, acquiring en route the characteristics of other innovating groups such as R and D, but, like the programmers, not required to undergo formal professional training. Therefore both the history of computer technology and the required characteristics of the computer technologist's innovative role would seem to make other industrial scientists appropriate reference groups at present, particularly for the systems analyst.

It seems likely that as the occupation develops, professional qualifications will become a necessity for those who wish to become programmers or systems analysts and the group will come even closer to the industrial scientific community. This could reinforce existing technical values and remove the group even further from the needs and values of computer users.

The influence of occupational literature on role definition

Reading professional journals is another means by which the specialist defines his occupational role and decides what he should be doing. If there is a wide gap between what a member of a specialist occupation is being required to do by his firm, what he sees his colleagues in other firms doing and what the journals he reads suggests he should be doing, then if he wishes to make a career in his specialism outside his present firm, he is likely to leave in order to secure a wider and more appropriate form of experience. A clearcut image of an occupation can be projected by those specialist journals which describe its activities. The British Computer Society until recently produced two journals: *Computer Journal*, with a reputation for very esoteric articles on technical computing subjects; and the *Computer Bulletin*, now no longer in existence, which took a broader position and published articles on a wide range of subjects seen as relevant to the work of computer specialists.

We argued earlier that the computer specialist defines his role very much in technical terms and does not recognise that it could, or should, have a content area concerned with human design aspects. That is, with the structuring of work activities associated with the system so as to increase job satisfaction and the implementation of systems in a manner that makes them acceptable to people. We have examined the titles of all articles published in the *Computer Bulletin* from 1969 to 1971 in order to ascertain how many of these were on non-technical subjects and whether any discussed our particular areas of interest – sociotechnical systems design and the human problems of implementation. Our results are as follows.

In thirty-six issues of the *Computer Bulletin*, containing some 114 articles, there were ten articles on the human aspects of computer usage and three of these were on the issue of privacy. The remaining seven included an article on problems of implementation, written by one of the authors of this book. The others were on the sociological consequences of automation, education for computer usage and the personnel problems associated with the use of computer specialists. There was no article on the human aspects of systems design. The *Computer Journal* contained no articles on computers and human beings. It can be seen that influences on occupational role definition which emanate from specialist computer journals reinforce the computer specialist in his definition of his occupational role in technical terms.

Occupational identification

Another factor which may influence a group to consider carefully the boundaries of its role is occupational identification. By this is meant the importance of the occupation for the individual in the sense that he intends to make it his career. This has been a source of frustration for many computer specialists because the early glamour and excitement of the specialism meant that many individuals became strongly committed to staying in it all their working lives. Yet today, the recent recession in computer sales, together with technological advances which are making some human programming activities redundant, mean that there is a promotion blockage at the top of many computer departments. There is a severe shortage of senior management positions in computer work and this is forcing many senior staff to move outside EDP into general management positions.

In our firms we found that 37 per cent of programmers and 27 per cent of systems analysts said that they would like to spend all their career in data-processing, while almost all programmers and systems analysts wished to stay in data-processing for the following five years. These attitudes seemed to show a grasp of labour market realities. Although there was great enthusiasm and satisfaction with computer work and many would have liked to stay in it all their lives, there was a recognition that for the majority career progression meant an eventual move into a general management job. It would therefore appear that long-term occupational identification is not at present a significant influence on the definition of the computer specialist's occupational role. It may become more influential when the profession settled down.

Differences in role definition and how these affect role behaviour

We have seen that the goals and needs of employers together with the values and experience of individuals in particular occupational roles play a major part in how these roles are defined. They determine the parameters of those sets of activities for which an individual believes he should accept responsibility when in a role and his philosophy on how these activities should be carried out. They determine also the expectations of those groups with whom he associates in his occupational role as regards what he should be doing and how he should be behaving. With many older occupations the role has become tightly specified on the basis of years of custom and practice and role holders and those who interact with them have the same expectations. With new occupations there may be differences between what the individual thinks he should be doing in his occupational role, what he thinks others expect of him, what others do expect of him and what *is* or should be expected of him if he is to meet the needs of his employing organisation. When this incongruence occurs the individual is likely to experience *role conflict*. That is he can find himself in a situation where different groups all have different expectations of him, or where the ideas of the major group with which he must interact conflict with his own definition of what he should be doing. This conflict can produce a stress situation for the individual which, if he has not the personal psychological resources to enable him to cope with it, will lead him to leave the job or, if he cannot do this, to a state of great psychological discomfort.

Frequently the situation is not as serious as this and, with new occupations in particular, the individual may suffer not so much from role conflict as from role ambiguity, a general uncertainty both on his part and on the part of others about what he should be doing, how he should be doing it and where his responsibilities begin and end. Companies do not always take enough care to ensure that there is compatibility between their expectations and those of the person in a particular occupation and this may cause problems from recruitment onwards. Many early computer systems suffered from a high labour turnover because computer specialists were attracted by jobs described in exciting development terms only to find, once the system had been designed and implemented, that the role changed from having a high research and development content to a more routine one of maintaining a successfully implemented system. Similarly, an organisation which has a strong feeling for good employee relationships, has built up a valuable reserve of goodwill in its staff and is unwilling to upset this, may find that it encounters difficulties if it employs the very technically orientated computer man who sees his occupation solely in terms of meeting technical objectives and has little interest in, or understanding of, the human needs of employees.

We have found that one of the problems which arises from the external recruitment of computer specialists is that they are likely to bring with them values very different from those existing within the organisation. This means that when new computer systems are being introduced there is likely to be a clash of values between the technical values of the computer men and the human relations values of the personnel department. In this kind of situation conflict may be less if computer specialists are obtained through internal recruitment, with the firm having the facilities to train its own computer staff. When people are drawn from the same backgrounds there is likely to be more of a consensus on the way in which problems should be handled and the manner in which people should be treated. Both Falcon Ltd and AEK had an internal recruitment policy of this kind and they appeared to avoid the difficult relationships between computer specialists and user departments which, on occasion, were found in our other firms.

How the systems analysts in our firms defined their role

When asked to describe their occupational role in terms of areas of responsibility our systems analysts gave similar answers. Table 6.4 shows that they provided us with a sequence of activities associated with the use of computer technology to solve business problems. They told us that they had first to define the problem and this involved collecting information. They next had to analyse the problem and arrive at a solution. After this they designed a new system of work based on the use of computer technology; they passed this

Table 6.4 Systems analysts' definition of their work (percentages)

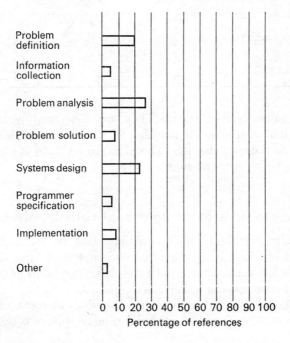

Percentage of references

system on to programmers so that it could be translated into a language which the computer could understand and instructions given to the computer, and they implemented the system. The three areas to which they referred design. There were surprisingly few references to implementation as a major activity even though all the other parts of the job would appear to be fruitless if they cannot be satisfactorily implemented.

The first interesting thing about the systems analysts in our firms is that they defined their job responsibilities in different ways. We have already seen that the Poulton systems analysts were much more likely to see the content of their jobs as purely technical or as technical and administrative than were the systems analysts in Falcon Ltd, and AEK who were more likely to perceive a human relations content in their work (see page 147). We have hypothesised that these differences in attitude reflected the differing recruitment policies of the three firms, for it will be remembered that Poultons tended to recruit its systems analysts straight from university.

All three groups were convinced that the nature of their work role made them view problems in a different manner from the managers and staff of other departments in the organisation. Table 6.5 shows this clearly.

Table 6.5

Question: Do other groups see things differently from you and your department?

	YES		NO	
	N	%	N	%
Poultons	15	88	2	12
Falcon Ltd	14	82	3	18
AEK	4	57	3	43

Typical comments received were: 'They are concerned with their own problems. We look at it from the outside.' 'Yes, our aims are different. Most users want to do a day-to-day job. The pressure on them is such that they cannot look for better ways to do it.'

Values as an element in occupational role definition

The way an individual defines his occupational role in a job such as systems analysis, which is relatively unstructured and contains a large discretionary element on how it is carried out, will be closely related to his personal values, to the values of those groups whose opinion influences him and to the values of the organisation for which he is working. By values we mean a set of beliefs or standards of excellence which enable us to make choices between possible ways of behaving. Our personal value position tells us that some things are right and desirable and that others are wrong and to be avoided. Our values give our life meaning and direct our behaviour, and we share our values with other individuals and groups who are similar to ourselves. We acquire our social values from the social atmosphere which surrounds us, from our families, our friends, our workmates (Sherif, 1963). Once acquired these values become internalised, they become a part of us and produce lasting social attitudes. It is this internalisation that prevents us from reacting in a random and unpredictable way when we meet new people and new situations.

In the same way as individual behaviour is influenced by personal value systems, so organisations have values which influence their business behaviour. The community in which a firm is located often makes a judgment on how these values are implemented by management and firms acquire local reputations for being ruthless, paternalistic, welfare-minded. The firm with a single powerful head who himself has a strong personal set of ethics which he uses to mould the thinking and behaviour of his managers will have a very coherent set of values. The early Quaker firms, for example, were run by men whose personal ethics gave them strong beliefs on how they should treat their employees and these beliefs were implemented at every level in the company. Most firms today do not behave in such a coherent manner and appear to have values which are not constant but are too often a response to pressures originating in the firm's environment. Thus a firm will argue that it can only look after the welfare of its employees when times are good and it is making a profit. There are also two sets of values which appear to influence a great deal of modern business behaviour. One is the technical ethic to which we have already referred. If a machine can do a better job than a man then let us use the machine. The second is represented by the phrase 'we are in business to make a profit'. This appears to be the excuse for a great deal of business behaviour which takes little account of employees' needs and interests.

What we are saying here is that because large firms today have confused and vacillating sets of values they are unlikely to provide much guidance to the individual specialist on how he should carry out his job responsibilities. Therefore, although his perceptions on how he should behave will be influenced by the strength of the firm's technical and profit ethics these will not provide him with guidance on how he is expected to behave in terms of personal relationships. Here he will have to rely on his own personal value system and the principal influence on his behaviour within the firm will be his peer group and the user groups with which his work activities bring him in contact. His responsiveness to the value systems of user groups will be a product of, first, his openness to communication from them, and second, the power of these groups in the sense that they can force him to take account of their ideas. An unwillingness to respond to user fears and wishes will create a situation full of ambiguity and uncertainty for the user department.

We asked our systems analysts what they considered were the qualities

which made a good systems analyst. Their answers are set out below (Table
6.6). An ability to deal with people and to communicate successfully with
them was given as an important quality by nearly a quarter of our systems
analysts, followed closely by a need for patience and thoroughness and intelligence.

Table 6.6 Systems analysts' assessment of the qualities their job required (percentages)

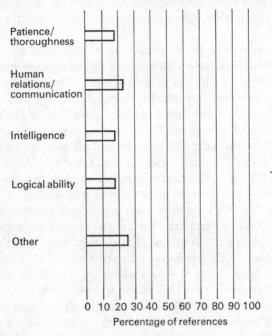

Other=personality characteristics such as motivation, ability to satisfy the client,
efficiency, creative and critical ability.

A second question on the kinds of skills required in their jobs produced a
similar pattern of answers. Management and human relations skills were seen
as most important with mental skills coming second.

Table 6.7 Skills required by systems analysts

	%
Management and human relations skills	36
Mental skills	32
Temperament	18
Technical and business knowledge	14

By having the right temperament systems analysts meant something very
similar to human relations skills. They spoke of an ability to exercise tact; to

avoid conflict, and to maintain good relationships. The same emphasis on temperament came in answer to the question 'What kinds of people make the best systems analysts?' It was stressed that anyone who was impatient or bad-tempered could not function successfully as a systems analyst. Temperamental defects were seen as more disabling in terms of job success than an absence of logical ability.

This emphasis on human relations ability and the need for a calm and equable temperament might be interpreted as showing that systems analysts had a considerable social awareness of what they were doing and of the need to meet human requirements when designing systems. Unfortunately this was not the case. Human relations skills were universally perceived in a manipulative way as an ability to persuade the user that he wanted the computer system, to communicate to him the nature of the system, to calm his fears and avoid trouble. In no sense was it meant as the ability to use democratic and participative processes, or an ability to identify the human results of systems, or an ability to design systems so as to increase job satisfaction.

We asked systems analysts to declare their value system on how their job should be carried out by asking them what methods they used to get innovation accepted. Again persuasion and education were the approaches most favoured in Poultons, Falcon Ltd and AEK.

Table 6.8 Methods for getting innovation accepted

	PROVE IT TO BE OF VALUE TO THEM		PERSUADE, EDUCATE		INVOLVE IN PLANNING AND DECISION TAKING		OTHER	
	N	%	N	%	N	%	N	%
Poultons	5	31	8	50	1	6	2	13
Falcon Ltd	4	24	7	41	4	24	2	11
AEK	–	–	5	71	2	29	–	–

It can be seen that few of the systems analysts in our firms believed in associating democratic processes with the introduction of major technical change. Answers were generally in terms of 'getting the other side to think as we do'.

> Do not blind with science. Discuss their problems; gain their confidence so that they are on your side when you present computer solutions.

> Joke them into something early on, then implement it.

> Persuade them that our system is well thought out and will work.

We were surprised that it was young men who were giving these kinds of answers for youth these days is very much identified with pressures for more democratic processes. We wondered if this reluctance to use a democratic approach was due to a desire to retain a clear command over their activities; to get the system in essentially as they had designed it without having to compromise the logic of their systems design by meeting user demands. We did find a relationship between educational background and a willingness to use manipulative or coercive methods for getting systems introduced, and it was the young men with university backgrounds and science degrees who were more likely to favour this kind of approach. These young men tended also to be the group which had been recruited from university straight into data-processing departments. They were in their first jobs and it would seem that they had not yet learnt either humanity or the political realities of industrial organisation. Individuals who were in favour of providing opportunities for user involvement in systems planning and design tended to have worked in a number of jobs in the firm and these previous jobs had not been associated with data-processing.

At the time this part of our research was carried out, the late 1960s and early 1970s, data-processing was on the crest of a wave. Senior management was still looking at this new technology uncritically and assuming that considerable financial benefits would accrue from its use. Because of this there was a tendency to receive computer specialists on their own terms and to accept as correct the way they were defining their occupational role responsibilities. Since then, the economic situation has hardened and computer systems as a form of investment are now more likely to be given very careful scrutiny. Inevitably this has meant a tightening of controls in many computer departments and a closer specification by top management of what they are to do. At the same time many user departments are now experienced in the use of computers and are no longer prepared to be the passive recipients of systems designed by computer specialists. They wish to avoid uncertainty by controlling their own environment. These new pressures may be influencing computer specialists to adopt a more democratic approach to their work. But we would need to carry out our survey again now in order to establish whether this change of attitude and practice has come about.

Many difficulties in our firms arose because of the way systems analysis was defined at the time the study was carried out. The systems analysts we interviewed had, as we have seen, a clear perception of their occupational role and the activities it comprised. They placed a great deal of emphasis on the design of logical systems and on persuading the user to accept these. They did not see implementation as a major skill area and they had little belief in the virtues of democracy and user involvement. For their part, user departments tended not to challenge this definition. They knew very little about systems design and what it entailed, and they had no clear role perceptions or role definitions to substitute for those of the systems analysts. Because of this they did not exert any pressure on the computer specialists to redefine their roles in order to meet users' interests more fully. Systems analysts, as a new occupational group, had an interest in defining their work territory in a way that suited their skills and interests and gave them maximum control over what they were doing. They had to establish a position of power and prestige in their companies if they were to be able to make claims for scarce resources such as money and influence. The user departments, for the most part, were on the wrong side in the power game. If they refused to accept new computer systems or questioned these too actively they were likely to be seen by top management as resistant to change.

Today the relationships between computer specialists and user departments appear to be moving on to a more equal footing, and the role expectations of each group are clearer and more in line. Communication is two-way rather than one-way, and because of this a consensus is growing as to the nature of the systems designer's role. However this consensus extends only to more discussions with the user on his needs and more involvement in planning and design of the system. It still does not include him as an equal partner in the systems design process and there is little recognition of the fact that effective systems design requires skills in the human as well as the technical part of the design process.

Ideology and power as factors influencing occupational role definition

An interesting feature of computer departments which influences their definition of their work responsibilities is the fact that there appears to be a dual ideology in existence. There are departments which see themselves in a powerful innovatory role, acting as a ginger group to the rest of the firm and pushing new methods and ideas out to the rest of the organisation. This kind of group will usually generate a great deal of uncertainty for those who are being pressured to innovate. In contrast other groups see themselves as 'helpers' rather than innovators. They have been described as the fact gatherers, the fact processers and the interpreters (Reid, 1968). We found these ideological differences in our case study firms. The Poulton and Grant systems analysts saw themselves in the ginger group role, whereas in Falcon Ltd, and AEK they appeared to fit more into the 'service' role. We have seen that these two last computer departments had a democratic committee structure to assist their decision-making and were more open to communication from user departments.

So far we have examined what have seemed to us to be some of the principal influences which led our systems analysts to define their occupational role as they did at the time of our study. We have indicated how their perceptions of their responsibilities influenced the way they behaved towards the departments into which they were introducing computer systems and, in turn, the amount of anxiety and uncertainty which they introduced into these user departments. We have suggested that there was a lack of congruence between the expectations of the systems analyst on the nature of the activities which should comprise his role and the expectations of user departments. This difficulty was exacerbated by the fact that user departments often had very unclear expectations. The role of the systems analyst was new to them and they were unsure what demands they could make of him or what constraints they could impose on his activities. We believe that this gap is now being reduced and that systems analysts are starting to redefine certain aspects of their occupational role on the basis of a changing and clearer set of user expectations. This process is being facilitated by a more equal balance of power and better communication between systems analysts and the staff of user departments when new systems are being introduced.

We have already discussed the issue of power as a factor influencing the decision process. It is also a factor affecting the way an occupational role is defined. All occupational roles are dynamic to a degree and changes in role definition may be a result of changes in the amount of power, discretion and responsibility permitted to the role-holder. Specialist occupational roles such as that of the computer systems analyst are particularly dynamic. They are constantly having to be redefined as their environment changes and they become subject to the influences of new technological developments, new ideas on organisational structure, changed expectations from groups which interact with the occupants of the role such as top management and user departments, and changed power relationships within the firm. Any occupational role is a product of the consensus between those performing the role and others who make demands or have expectations of this group, on what activities should be included in the role and how these should be carried out. If these demands and expectations alter over time for any reason then there will be pressure for the occupational role to be redefined. When more than one group wishes to have responsibility for a set of tasks there are likely to be power battles over where the boundaries of each occupation are to be drawn. The history of industrial relations is full of demarcation disputes on this issue.

We have defined power as access to scarce resources for which a number of groups may be competing, or which other groups need but do not have. We argue that technical knowledge is an important resource in this respect and one which may enable the expert to create a situation of dependency with others who require his skills. But, over time, relevant technological knowledge may shift from one group to another. This shift may be from one specialist to another specialist, as has happened with programmers and systems analysts. It may also be from the specialist to the non-specialist and as line management and the staff of user departments acquire a knowledge of how to design and use computer systems so their dependency on the computer specialist will decrease and his power position be reduced in consequence. This shift in knowledge to user groups is happening, but only slowly. In the meantime user departments which believe themselves to be threatened by new computer systems may have to resort to other strategies. Internal cohesion is one method of protection. If a group believes that it is threatened by another group then it is likely to draw together, to show a collective identity and to introduce group norms directed at emphasising group unity and solidarity. In addition it may attempt to demonstrate that its skills are still relevant and to throw doubts on the competence of the computer specialists. Again, it may try to influence the firm to be cautious and to move slowly when introducing these new forms of technological change. In the next chapter we examine the relationships between computer specialists and users in one of our case study firms.

7

Computer specialists and user departments - role definition and expectation[*]

In the thirteen or fourteen years in which computers have been in general use commercially there have been dramatic changes in computer technology. These changes have kept user task environments in a state of flux and uncertainty, and have led to the generation of considerable anxiety in line departments when the introduction of a computer system is first suggested. Because of the differences in the attitudes, values and interests of line departments and computer specialists one may expect their decision-making deliberations to be characterised by tension, conflict and misunderstanding. Pondy (1967) has noted that joint decision-making activities between specialist groups and managers are more likely to be characterised by bargaining than problem-solving. Lynton (1969) notes that such bargaining is likely to include the 'careful rationing of information and its deliberate distortion; rigid, formal, circumscribed relations: and suspicion, hostility and disassociation. Walton (1965) talks of minimum disclosures, manipulation, defensiveness and distrust.

In this chapter we argue that a source of conflict and uncertainty within the computer decision-making and implementation situation is the fact that two very different kinds of groups have to interact together. On the one hand are the experts, who have responsibilities and values different from those of the line man. These, as we saw in chapter 6, are a product of the network of roles and tasks which the expert creates in order to perform his organisational duties and also a result of his professional training and personal value orienta-

[*] This chapter is based on an investigation made by Krystyna Weinstein (1971) a member of the team which carried out the research.

tion. The computer expert must have a vested interest in innovation and change, for the continuation of his department and job depends upon his ability to introduce change. On the other hand there are the line managers who are likely to be more interested, at least in the short term, in meeting production targets and deadlines and in seeing that their departments have a day-to-day efficiency rather than a major capacity for change. The interests of these two groups are somewhat different and these differences will become apparent once they confront each other within a decision-making situation.

The computer specialist may also believe that his relationships with line managers are complicated by the fact that he is perceived as having no line authority of his own and therefore being dependent on the approval of managers who possess this formal authority. Like other specialists, computer experts have traditionally been regarded as staff officers; they give advice but not orders. Pettigrew (1968) and Mumford and Ward (1966) have argued that this staff–line dichotomy is no longer descriptively nor analytically useful. The nature of the computer specialist–manager relationship is likely to be characterised as much by persuasion, consultation and influence as it is by advice. In many situations the superior power of the computer specialist, which, as we have seen, is a product of his skill and knowledge and the influence he can exert on top management because of this, may mean that he is in a position to tell the line manager what he is to do.

Mumford and Ward (1966) argue that the computer specialist performs a role very different from that of the traditional staff advisor. 'These specialists are not there merely to assist line management to perform its existing duties in a more efficient way. . . . Their task is something more basic and fundamental to the commercial success of the enterprise. . . . Their slogan could be said to be "if it works, it's obsolescent" '. The changes computer specialists recommend may alter 'the functions of management and perhaps eliminate some management positions altogether. Therefore, unlike normal staff advisors, the new specialists represent a threat to the jobs and power positions of many line managers.'

In any innovative decision process involving executives and technical experts, such as computer specialists, the issues likely to arise will have to do with the relative claims which each side makes for its knowledge and skill to be regarded as important resources. The computer group is likely to frame its demands in the form of a need for increased recognition of the importance of technical information as a business resource. This will increase its personal or group standing as the source and controller of such information. Yet, such a request may be interpreted by executive managers as a demand for quasi-elite status, which in fact it may be. In this way new political action is generated and the existing distribution of power within the organisation is endangered.

While in some innovation situations there will be jockeyings for power, influence and control between computer specialists and future system users, in others a disbalance in power will exist from the start and be maintained throughout the decision and implementation processes. If in this kind of situation the user group does not fully understand the role of the innovator group and does not know what it can legitimately expect in terms of behaviour from this group, then a great deal of uncertainty, anxiety and conflict is likely. How the role of the computer specialist is perceived is therefore a critical factor in the generation of uncertainty. If this role is clear then users will be able to bargain over which group has responsibility for different aspects of the innovation process. If the role is unclear and ambiguous and, in addition, the user department has a low degree of power, then bargaining may be replaced by covert resistance or by attempts at dissociation from an unpleasant situation. But, as we indicated at the end of the last chapter, occupational roles are dynamic, and the achievement of role clarity may not always be easy. It is likely to be particularly difficult with new occupational groups such as computer specialists who may not themselves have defined their own role with any precision. A user department having its first encounter with computer specialists will have great difficulty in comprehending the nature and extent of the innovator's role.

Problems of occupational role definition

In the past, many of the studies made of different roles in society have appeared to assume that roles are fixed and that there is a consensus on the set of activities which go with any given role. Linton, for example, has described a role as the 'attitudes, values and behaviour ascribed by society to any and all persons occupying this status' (1965). This definition seems to imply that someone has decreed what a given role shall be and what the holder of it shall do, and that this prescription is universally accepted.

A more realistic perception of the occupational role is to view it as a set of tasks delineated as a result of a continual process of negotiation between groups which interact together. Allocation of these tasks at any time will be strongly influenced by group objectives, and these in turn may be related to attempts to obtain scarce resources such as money, status or power. Therefore, although the core of an occupational role may be fairly static in that it contains a set of activities which are not challenged by others, many roles will have mobile boundaries.

Organisations often seem to pretend that the network of roles and activities which they comprise are clear, fixed and definable. This is shown by the very precise job descriptions which many firms keep stored away in the files of their personnel departments. Nevertheless, in reality, there may be little similarity between a formal job specification and the activities actually carried out by the person holding that occupational role. Role definition in situations where jobs are not tightly structured is likely to resemble the processes described in our programmer–systems analyst case study with tasks allocated on the basis of a set of reciprocal responses between groups that work together. One group may challenge the activities of another and seek to take some of these over;

this compels the first group to rethink and modify its activities in order to maintain a working relationship with the second group. In this way roles emerge through a process of what Turner (1962) calls 'feedback testing transactions'.

History plays a part in the definition of most occupational roles. The holder of an occupational role with a tradition behind it will have to respond to expectations, built up over time, of what he should be doing and where his responsibilities lie. The content of these expectations, as Katz and Kahn (1966) point out 'may include preferences with respect to specific acts and personal styles; they may deal with what the person should do, what kind of a person he should be, what he should think or believe; and how he should relate to others'. A person in an occupational role with a long history may find that his potential for remoulding the role is severely limited.

However, there will always be an element of unpredictability in the ways roles develop. In addition to external pressures such as changes in technology or client needs, the behaviour of the individual occupying the role must affect the way the role develops. The future consequences of present actions will play a part in shaping the route the occupational role takes. The ability of a role occupant to influence the development of his job in a particular direction will depend on his skill in predicting correctly the consequences of his actions. We are talking here of occupational roles with a reasonably large discretionary content, and these are more likely to be found in management and specialist groups than in more tightly structured activities.

Special problems arise for the individual in an occupational role which not only has a large discretionary element but is a boundary role in the sense that the occupant has to interact with several other groups. If these groups do not have similar expectations of what he should be doing, then he has to make a number of difficult decisions. These concern how he reconciles these differing and perhaps conflicting expectations, and, if he cannot do this, which set of expectations he chooses to follow or give priority to. An example of a supervisory role in which this dilemma occurs is that of the coal face deputy. The deputy has a statutory responsibility for safety, yet at the same time he can be pressured by his management for higher production and by his men to ensure that their earnings are protected. On occasion he can only achieve safety at the expense of production and vice versa. He then has an extremely difficult decision on which set of expectations to meet (Scott *et al.*, 1965).

Problems of defining new occupational roles

Where an occupational role is either entirely new or new to a particular organisation then there are additional problems about its definition. It has no history, therefore no set of rights and duties has, over time, been associated with it (Linton, 1965). There are no precedents to guide the behaviour of the individual in the role. New roles are most likely to emerge during periods of innovation when new technologies and new forms of organisation throw up a demand for new sets of skills. People moving into these new roles are then involved more in 'role-making' than 'role-taking' (Turner, 1962). The situation is complicated even further if the new role is operating in a constantly changing environment and has to be continually modified to fit with the demands of this environment. It then becomes very difficult to incorporate any stability into the role or to develop a set of clear cut expectations on the part of the occupant of the role, and those groups interacting with him, on what his responsibilities are. The role of industrial consultant would come into this category. Most people in industry would see him as some kind of problem solver and 'helper' and he is likely to define his own role in this way but it is difficult to make this definition more specific (Mumford, 1972). This can only be done in terms of each situation he enters after his clients make clear to him the nature of the problems they want him to solve and he makes clear to them how he is proposing to solve these problems.

We have seen that many occupational role relationships are a product of negotiation over time between interacting groups. This poses the question of what happens when a firm is faced with a totally unfamiliar problem in which a new and unknown group has to cooperate with groups which are long established in the firm and have developed mutual agreements on the sharing out of tasks and responsibilities. Such a situation is likely to be fraught with uncertainty due to a lack of knowledge of what can legitimately be expected from the new group. A process of bargaining for task areas will certainly take place, but this will be clouded by a lack of understanding of the nature of the tasks available for bargaining. In this kind of situation the incumbent of a new occupational role, for example a systems analyst, is likely to attempt to structure his role in terms of his own predispositions, knowledge and interest. In chapter 5 we set out some of the factors which would influence his role structuring. He may wish to accept responsibility for certain sets of activities but not for others, and he may hope to persuade other groups to take over unwanted task areas. A problem for him will be to establish first, that these groups have an interest in these work activities and, second, that they are competent to undertake them.

In our case study we show how a lack of competence on the part of user departments forced the computer specialists to redefine their role in wider and more executive terms than they originally intended. In this kind of novel and unclear situation a desire for power and influence may be less important in the allocation of tasks and responsibilities than the evaluation groups make of their competence to assume particular responsibilities. The success of this evaluation and task-sharing will depend on an ability to spell out in detail the nature of the problem to be tackled and the processes of problem solution. Another factor affecting task-sharing will be the urgency of the solution and the time available to allow inexperienced groups to gain new competences (Goode, 1960).

In a situation with new problems and new roles the process of securing agreement on the allocation of tasks is certain to be stressful and painful, and there will be a great deal of taking up, dropping and transferring of task activities as each group attempts to define its activities in terms of its competence, needs and objectives. Kahn *et al.* (1964) point out that when this process of allocation is difficult to resolve conflict is liable to break out between the groups. Some individuals and groups will react to this conflict with aggression and make attempts to stake their claims forcibly, others will respond by attempting to withdraw from the situation and leave others to take the decisions.

Role expectations

If we accept that roles are not fixed but changing it becomes very difficult to define an occupational role and describe its content. It can be argued that this can only be done through examining the expectations of the role occupant on what he should be doing and the expectations of those groups with which he interacts. If these expectations are similar to each other there is no problem and at any given moment it is possible to describe with some clarity the prescribed responsibilities and discretionary areas which make up this occupational role.* But if these sets of expectations differ greatly then the problem is more difficult and this lack of agreement may be an indication of the complexity or newness of the role.

A person moving into a new occupational role, or into a role which is new to his organisation, has to set about defining what his role should be and moulding the expectations of others as to what he should be doing. He will have certain guidelines here, and Goffman (1959) points out that 'if an incumbent takes on a task which is new to him but also unestablished in society, or if he attempts to change the light in which his task is viewed, he is likely to find that there are several well established paths among which he must choose'. For example, we have argued earlier that computer specialists tend to choose between assuming a 'helper' role in an organisation, seeing themselves as the servants of line management, or taking a 'ginger role'. believing that they have a responsibility to persuade the organisation to accept more innovation. Which of these roles a computer department adopts will depend on the personality and interests of the manager and his staff and their weighing up of the needs of the situation in which they are operating. Nevertheless a computer department will be constrained in its approach to problem-solving by the expectations and competences of other groups in the firm. A 'helper' role is difficult to assume if line management are apathetic about change and unable to define their own problems. Similarly, a 'ginger role' may prove unacceptable in a firm where line managers are used to handling innovation

---

* Wilfred Brown (1960) has described the prescribed part of a job as those areas for which an individual has been given clearcut responsibility. The discretionary aspects of a job are where an individual can make choices; for example, whether to plan his work in one way or another, whether to develop his responsibilities in one direction or another.

and recognise clearly the development aspects of their own managerial roles.

An individual entering a new occupation has preconceived ideas on what he should be doing, derived from his training and his contacts with others in the same occupation, but he will be constantly looking for clues from others in the firm in order to identify their expectations of his role. If he discovers that there is a gap between his role definition and the expectations of others on what his activities should be, then, if he is wise, he will try to reduce this gap. He may do this in a number of ways: (1) by persuading and educating the other group to modify their expectations and come into line with his own; (2) if he depends on their cooperation or sees himself as providing them with a service, by adjusting his behaviour so that it comes closer to group expectations; (3) by educating and adapting simultaneously. Unfortunately, members of new specialisms, because they are usually seeking status, structure and security, sometimes appear to cut themselves off from feedback on how others see them and wish them to behave. When this happens relationships become ambiguous and confused; there are no clear expectations in the situation, and conflict is prone to develop between new and established groups. Figures 7.1 and 7.2 show the relationships likely to occur when computer specialists and the users with whom they relate have different kinds of expectations.

Some ideas will always exist in people's minds on new roles and how they should be enacted, even though these expectations may be limited and vague. But the occupant of a new role has a problem when no consensus exists and each group with whom he interacts has different expectations of him. A study of the role of American school superintendents made by Gross *et al.* (1958) found, for example, that people who interacted with the school superintendent all had very different expectations of the nature of his responsibilities. Homans (1965) has pointed out that a factor leading to these differences is the position held by the other individual. If he is a manager he may have a very different set of expectations than if he is a clerk or factory worker or a union official. When there are different sets of expectations concerning a particular occupational role, the role occupant is likely to have an unstable relationship with other groups because he is continually presented with conflicting priorities. Kahn *et al.* (1964) have identified four different kinds of role conflict. These are

1 *Intrasender conflict.* This is when a group or individual with whom a role occupant interacts has conflicting expectations of what the role occupant should do. Our colliery deputy, required by his management to look after safety and production at one and the same time is a good example here.
2 *Intersender conflict.* This is when different groups with which the role occupant interacts have different expectations of him. For example, the expectations of top management on what a computer specialist should be doing may conflict with what the manager of a department where he is introducing a new system thinks he should be doing.

Fig. 7.1 Agreement on role definition

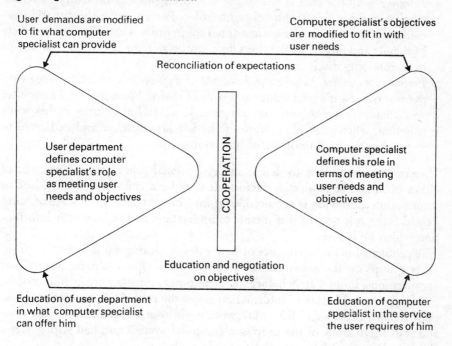

User demands are modified
to fit what computer
specialist can provide

Computer specialist's objectives
are modified to fit in with
user needs

Reconciliation of expectations

User department
defines computer
specialist's role
as meeting user
needs and objectives

COOPERATION

Computer specialist
defines his role in
terms of meeting
user needs and
objectives

Education and negotiation
on objectives

Education of user department
in what computer specialist
can offer him

Education of computer
specialist in the service
the user requires of him

Fig. 7.2 Role conflict

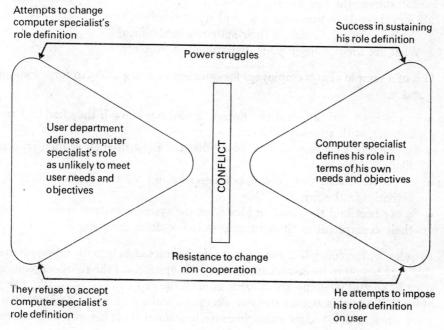

Attempts to change
computer specialist's
role definition

Success in sustaining
his role definition

Power struggles

User department
defines computer
specialist's role
as unlikely to meet
user needs and
objectives

CONFLICT

Computer specialist
defines his role in
terms of his own
needs and objectives

Resistance to change
non cooperation

They refuse to accept
computer specialist's
role definition

He attempts to impose
his role definition
on user

3 *Interrole conflict.* This is where the role occupant has difficulty in reconciling a given role with his other life roles. For example, the situation encountered in an Irish factory in a small community where the foremen were
husbands and fathers of the shop floor workers. They had serious problems
in maintaining discipline.
4 *Person-role conflict.* Where the demands of a given role are at variance with
the role occupant's own values and code of ethics. For example, a computer
specialist may be required to act as a hard, technical man in his work
situation, whereas in his private life he is a lay preacher and dedicated to
improving the quality of life of his fellows.

Uncertainty about how to define an occupational role may be a product of
either objective or subjective factors. It may be a consequence of a lack of
information as to what is required, or it may be a psychological state of mind
in which the role occupant is unable to understand and organise the information which he is given.

A problem for the occupants of new roles is picking up information from
other groups on the expectations of these groups. For example, in our four
organisations, immediately before their computer systems were introduced, we
found a remarkable lack of information about the role of the systems analysts,
even though these specialists were working in user departments at the time
and staff there knew of the proposed computer system and had strong views
about it. We asked four questions of user staff:

1 Have you had any contact with the systems analysts or other computer
staff during the past twelve months?
2 Does their department have a good reputation?
3 Can you describe some of their activities in the firm?
4 What are they doing in your department at present?

Out of a sample of 390 employees for questions 1 and 4 and 519 for questions
2 and 3:

- 71 per cent said they had no contact, or did not know if they had had any
contact, with systems analysts.
- 61 per cent said they had no idea what the department's reputation was
like.
- 70 per cent said they could not describe any of the systems analysts'
activities in the firm.
- 89 per cent said they had no idea what the systems analysts were doing in
their department or thought they were not doing anything.

In roles where there is a requirement to interact closely with other groups
there are bound to be certain areas where the opinions of the role occupant on
what he should be doing are at variance with the expectations of collaborating
groups. When this occurs the role occupant will suffer from strain; that is,
from stress which occurs either because he cannot meet his own and others'

expectations of what he should be doing or because he has to choose between courses of action which will suit the interests of one group but conflict with those of another. A great deal of energy is used by the occupants of such roles in trying to resolve these differences between their role expectations and those of others. Goode (1960), for example says, 'not only is role strain a normal experience for the individual, but since the individual processes of reducing role strain determine the allocation of role performance to the social institutions, the total balances and imbalances of role strains create whatever stability the social structure possesses'. Buckley (1967) points out that role modifications that are made by role occupants usually represent a working compromise 'between the structural demands of others and the requirements of one's own purpose and sentiments'.

Specialist groups

Specialists such as computer personnel are very prone to suffer from role strain and uncertainty and this arises because the organisation for which they work often lacks a clear understanding of how to use their skills. Top management may not provide its specialists with a clear picture of the tasks they are expected to perform, and indeed the nature of the specialist role may not be thought through in any meaningful way. Yet the specialist, if he is to operate effectively in an organisation must ascertain what assumptions others are making about his area of responsibility and establish the extent to which these coincide with his own role definition. If he does this and finds a discrepancy he must work out how he is going to reconcile these divergent expectations. This is particularly the case with specialists who have the function of introducing major technical change. Mumford and Ward (1968) suggest that a major source of difficulty for computer specialists is the fact that their survival as a viable group depends on their ability to press forward with innovation. In contrast, line managers may give their production responsibilities top priority, because their performance is evaluated in terms of their ability to meet production targets and delivery dates. A disturbance to a well run production department because of some risky new computer system may be viewed by staff there with great doubt and scepticism. They will have to be convinced that the activities of the computer specialists are in line with, rather than against their own interests and those of their department. When a lack of confidence exists between line management and computer specialists this can show itself in the kinds of interpretation each group places on the behaviour of the other. For example, Argyris (1971) found that whereas a computer specialist group saw itself as innovative, creative and rational, managers thought it showed the opposite tendencies.

When specialists belong to an external group sent into a local firm by the head office of a parent company their difficulties may be greater still, and there may be even more confusion over the interpretation of their role. The specialist entering such an environment will have few institutionalised guide lines to follow and, as Gouldner (1957) points out 'the more unstructured the situation, the more dependent he is on personal contacts and friendships'. These provide a means for him to understand expectations and mould his role to meet these. A difficulty for the outsider is that he often has less identification with local problems and needs than the internal man, yet the amount of involvement he feels will determine the role he plays *vis-à-vis* other groups in the situation. We have shown in an earlier chapter how computer specialists recruited within a firm were more identified with the problems and needs of line departments than were externally recruited computer specialists. These tended to take a tougher, more technical approach to their work and to have expectations that they could make others accept their technical values.

Questions illustrated by our next case study

An interesting question for the researcher examining a new occupational role is posed by Gross. He asks (Gross *et al.*, 1958): 'Is the extent to which there is consensus on an occupational role an important dimension affecting the functioning of a social system?' Our case study data illustrate a situation where this consensus did not exist and had to be arrived at through a slow and painful process of adjustment: formal role definitions, where these existed at all, were too imprecise to act as a guide to the role occupants. Role definitions were largely left to the role occupants themselves and roles were altered when it became clear that certain groups had competence for some activities but not others. Before this reconciliation occurred there was considerable misunderstanding on how responsibilities were being, and should be, allocated.

In the case study we examine the relationships between users and computer personnel during the planning and implementation of a new computer system and show how a lack of specificity in occupational role definitions and an unclear allocation of responsibilities for different aspects of the change process affected the successful management of this change. The study will demonstrate that occupational roles are dynamic; that the boundaries between groups that interact with each other shift and alter, and that this is especially the case when a major change is taking place. Particular problems occur with new groups for which there is an absence of precise knowledge on the nature of their responsibilities. These problems are exacerbated if these new groups are concerned with the introduction of high risk innovation. They may seek to protect themselves by avoiding task areas which they regard as threatening and by attempting to push these within the task boundaries of some other group.

A case study of innovator–user relationships

Our case study firm was the subsidiary of a large, extremely diversified, company and had a labour force of around 2,000. It had a strong company culture and operated in a flexible manner without tight job descriptions or even a clearly documented company structure. The Company Secretary remarked, 'People do not know what the structure of the firm is, but they have an idea. . . . They do know who they work for.'

In 1965 the firm became interested in the possibilities of computer systems and formed a committee to examine how this new technology could be used, when it should be used and what steps were required to prepare the firm for computerisation. This committee met once a month and part of its brief was to examine some of the firm's existing manual systems in order to establish the extent to which these would require 'cleaning up' before they could be considered for computerisation. Even in these early days there was an awareness of the need to keep staff informed of future developments and a note in the minutes states: 'We intend to pass on, by means of lectures, information to all staff concerned in the working of the new system.'

In 1967 an O and M analyst was appointed and given responsibility for examining existing work systems with a view to their future computerisation. He was assisted in this exercise by six 'liaison men' who were loaned to him, part-time, by the various production departments.

The Managing Director explained the firm's interest in computer systems as follows:

Our volume of business had grown to such an extent that it was no longer practical to run the business with a manual system. This firm has always been progressive and it had become clear to both our parent company and ourselves that we could no longer efficiently control our production using our old system. . . . All the managers here were agreed on this decision.

Another Board member commented:

Our scale of operations was such that we needed better control of our money . . . so much of it was locked up in stock. We discussed the whole thing informally with the other managers.

In 1968 the Managing Director went to see the parent company to discuss the possibility of his own firm purchasing a computer. The parent company, however, already had one computer in operation, and it was decided that before individual firms made their own local plans a group policy on computerisation should be developed. As a result of this decision a centralised computer service department was set up and located at the parent company to advise unit firms who wished to use computers. This centralised unit was given the following terms of reference:

1 the operation and development of existing computer installations;
2 the selection and recommendation for installation or development of new or replacement computers and associated equipment;
3 the supervision and recruitment of staff within agreed establishments;
4 the training of computer staff;
5 the training of user company staff.

Here was the first attempt at a role definition for Central Computer Services and at this moment in time its role was seen as advisory. In July 1968 the Managing Director of our case study firm sent his departmental managers a memo stating that two systems analysts from the parent company would be visiting the firm for one month to undertake a preliminary feasibility study for a new stock control system. The two systems analysts carried out their work during the summer and early autumn, though they had some difficulty in ascertaining exactly how manual systems operated in the firm as opposed to how staff believed or said that they operated. Nevertheless by late autumn they felt that they had a reasonable grasp of the stock control situation.

From the time of their arrival the presence of systems analysts from Central Computer Services generated a great deal of uncertainty. It led to an undercurrent of questions and doubts about the whole computer venture. Two questions repeatedly asked were: first, 'Why do we need a computer?', a question posed primarily by departmental managers and section heads; and secondly, 'What can a computer do for us?', posed by employees at every level in the firm. These two logical questions were accompanied by remarks about the need to tidy up manual systems first and the suggestion, frequently made, that if this were done there might be no need to introduce a computer. Other more specific questions posed at this time were: What is a computer? What will it do here? How will it solve our problems? Why are we having it? Do we really need it, if we get our manual systems in order? How will it affect individual departments? What will each manager have to do? How will his job and position be changed by it? What will be its effects on the company? Who will be affected by it?

This uncertainty did not mean that there was resistance to the idea of a computer system.

Despite the questions, in these early days there was an optimism that all would ultimately go smoothly. The stock control department saw their work becoming more interesting; they believed that much of the drudgery would be taken out of their work and that their present work load would be eased. These favourable first impressions pervaded the company even though at that stage (December 1968) no official information about the computer system had been given to staff as a whole and information was derived from rumour and gossip.

The feasibility study

The feasibility study report for a new stock control system was given to the Board in December 1968. It outlined the work to be transferred to the computer and its authors clearly recognised the need to deal with the uncertainty of a new system for it contained some statements about communication and education, as follows:

As computers are relatively new, it is essential that all levels of staff are educated to an awareness of the requirements and implications of this new technology and any possible fear of the unknown replaced by an appreciation of the advantages which can be obtained by the full use of an integrated computer system. In an endeavour to achieve this, emphasis has been placed on the important aspect of personnel communications and education in the implementation of this project.

Thus at this early stage the computer specialists were pointing out the necessity for communication and education policies as a means for reducing anxiety, and they stated in the document that they saw this as their responsibility.

The feasibility study also recognised the importance of ensuring that the user department was identified with the proposed change, for a section dealt in detail with the education of company personnel. It mentioned at some length the need to secure the commitment and enthusiasm of all staff, and stressed how resistance could lead to a lack of cooperation. It continued:

> A key group which must be committed to the new system is the company management. If they have doubts or gaps in their knowledge, they will inevitably pass on this uncertainty to their staff, and will be unable to give such reassurance and information which is essential during a period of major change.

The approach proposed was as follows:

GENERAL INFORMATION. Information of a general nature on computers and their application should be given to as many of the personnel as possible by means of selected films, lectures and group discussions.

DIRECTORS AND MANAGERS. The management should be totally involved with the implementation of the proposed system if it is to be successful, and they should attend an executive seminar provided by the equipment manufacturer.

DEPARTMENT HEADS. Computer appreciation courses should be attended by department heads to ensure that they have a knowledge which enables them to deal with the implementation of the system and the problems which will arise within their departments because of the change.

O AND M DEPARTMENT. One of the major tasks to be undertaken for the introduction of computer facilities will be the establishment of suitable procedures and disciplines within the various departments of the firm. . . . To ensure that these procedures are followed, arrangements should be made to confirm that the person who heads the O and M department is of managerial status and that he reports to the board. The O and M manager would be required to write specifications for work which is proposed for computer operation and he would require support by staff who can carry out detailed investigations into company procedures. . . . It is proposed that the existing complement of six be continued, and reviewed when the actual demand becomes more certain in approximately six months. Advantages would be obtained from locating the firm's O and M team next to the systems analysts.

Thus excellent plans were formulated at this stage of computer introduction and they were directed at securing the confidence and competence of user personnel. It is, however, noticeable that they came from the parent company systems analysts and not from the management of our case study firm. Local management had not yet become involved in the planning processes. The time scale for systems design, programming and implementation is shown in Fig. 7.3.

Fig. 7.3 Time scale for systems design programming and implementation

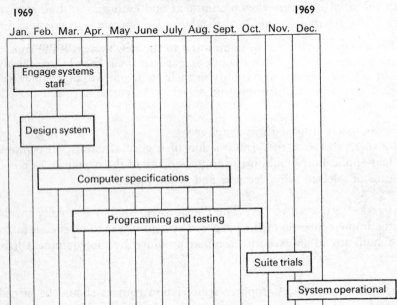

The feasibility study was presented to the case study firm in December 1968. It was discussed at a Board meeting in January and formally accepted by the Managing Director.

Line management uncertainty

At this point there was a great deal of uncertainty over the aims and objectives of the proposed computer system and uncertainty over whether there was a need for a computer based stock control system at all. There was also some scepticism over whether the firm could assimilate such a large change in such a short time. This uncertainty seems typical of new computer systems which have a tendency to start off in an atmosphere of doubt and hope on the part of user departments. This has been the experience of the authors and of many other observers. For example Mann and Neff (1961) point out:

> It is not at all uncommon for top management to decide to go ahead with a change that has been thought through and proposed by a small sub-group in the organisation, without any real understanding about the long run, indirect and marginal implications of the change proposed. . . . It is frequently difficult for top management to foresee the full implications that one of their direction-setting decisions will have.

The system was to be planned and implemented between January and December 1969. Systems analysts were to be engaged between January and April. These would form a new 'Computer Department' which would be located in our case study firm and assume responsibility for designing the computer based system; but which would be part of the parent company's Central Computer Services and would act only in an advisory capacity to local management. In association with this group the 'local' O and M Department was to check that 'suitable procedures and disciplines existed within the firm's departments' to ensure that 'clean' and not 'dirty' manual systems were being computerised.* The outline plan for computerisation was to be continued in a company procedure manual which would be compiled and distributed as soon as possible.

Early communications increase uncertainty

Towards the end of January the manufacturers of the selected computer were invited to the firm to give a presentation to all directors, managers and section heads. This session consisted of five half-hour presentations on computer hardware, and on the advantage of having a computer and the information it could provide. If anything, this increased managerial uncertainty for, so managers maintained, the demonstration was too technical, too complex and too abstract. At the end of the presentation there were still considerable doubts on why the firm needed a computer and on what it would contribute to company efficiency.

Two weeks after the manufacturer's presentation the Managing Director held a meeting with departmental representatives, to tell them, and through them all other employees, about the decision to have a computer. At this meeting he gave assurances of 'no redundancy' but pointed out that the new systems would require some retraining of staff and the reallocation of staff to other duties. He ended the meeting with some stern words:

> I ask everybody to do all they can to make this installation successful as without it the company has now reached such a size that we cannot operate effectively. Everybody in the works and offices must comply with the systems laid down, and very severe action will be taken against people who fail to comply with the necessary disciplines.

* A 'dirty' manual system is one in which there are *ad hoc* methods of working, which contains many inaccuracies and which is poorly controlled.

This last remark, although intended to ensure that there would be conformity to the demands of the new system, added anxiety to uncertainty. Staff now felt that they did not know what would be expected of them but, whatever it was, they might be punished if they could not do it.

The role of the O and M Department

In February a new O and M Manager was appointed from within the company. His terms of reference included:

> To assist with processing requirements for the phases of the computer installation . . . to control the O and M team, who would be responsible for systems as outlined in the O and M manual . . . to assist with the development of new systems in the future . . . this would involve disciplining systems through department heads . . . the collection of data prior to processing by group computer services . . . and generally be responsible for all aspects of systems, including *inter alia* their formulation, maintenance and discipline.

The rest of the O and M Department still consisted of the six 'liaison' men from various departments throughout the Company. These men were now to spend part of their time in O and M and the rest in their own departments, doing their normal work. The O and M Manager's terms of reference did not, however, provide any guidance on how he was to relate to the new computer department which had been set up. Nor did they clarify the relationship between the O and M Department and line management. The O and M officers now began to have problems arising from dual allegiance; an unclear definition of their own role, and unclear perceptions of how they should relate to the Computer Department. The O and M Manager attempted to clarify his position by defining his own role. He said, 'I see myself as a link man on the systems side, between departments and the existing O and M manual, and the computer staff . . . we as a department will do most of the firm's systems work and the computer staff will merely computerise it.' Unfortunately the O and M Manager's definition of his role did not fit that of the director to whom he was responsible. The latter was of the opinion that the Computer Department and the computer manufacturer would between them take responsibility for systems work with O and M assuming only a minor role.

From an early stage, therefore, there was a great deal of role ambiguity in the situation with conflicting expectations of the roles various groups would play in planning and designing the new systems. There were also differences between the responsibilities role incumbents thought they should assume and those that others were prepared to allocate to them. These conflicting expectations persisted as planning got under way. The manager of the parent company's Central Computer Services Department took the view that local O and M staff should do most of the donkey work of preparing the new systems, while the systems staff merely rationalised these systems for the computer. The local firm's Board, in contrast, saw systems analysis and design as the responsibility of Central Computer Services and the computer manufacturer.

There were also differing expectations on who was to take responsibility for informing and educating staff in the new system. It will be remembered that training had been specified as coming within the terms of reference of the parent company Computer Services Manager. However the local O and M Director told his O and M Manager that communication and education was his responsibility. Time was passing and there was clearly a need for someone to assume responsibility for communication and training as the last formal communication from management to staff concerning the new computer system had taken place three months earlier. But, because of the confusion over local versus centralised responsibility, little was done.

Other problems were occurring within the O and M department because of an absence of clear role definitions and an understanding of where the responsibilities of groups and individuals lay. The O and M Manager was dependent on his liaison officers for O and M work and these men were seconded to his department for two or three mornings a week. This was a somewhat haphazard arrangement and there was uncertainty among the liaison officers as to which manager had prior claim on time. A further source of uncertainty was whether they were merely departmental representatives or had any authority to make decisions.

The role of the Computer Manager

In the meantime the parent company's Central Computer Services had been advertising for a manager to run the local firm's new computer department. The manager was appointed in early April and his first two systems analysts arrived in May and June respectively. Immediately there was uncertainty over the respective roles of the Computer Department and O and M. The Computer Manager came, as he put it, expecting to 'run the show'. He was quite clear in his mind that he had an 'executive' role, that it was his staff who were designing the computer system and that this would merely be presented to local management for approval. Unfortunately the local group had quite different expectations of his responsibilities and his attitude did not fit in with the advisory role which had been adopted by his own manager in Central Computer Services.

The anxieties of local management

As the design of the system progressed the management of the local firm became increasingly uncertain about the role they were to play in planning and implementing the new stock control systems. The questions now being asked were much more specific and detailed than those of a few months earlier. There was frustration that there had been so much emphasis on 'communicating with staff' in early policy documents and yet there now seemed to be no one with responsibility for providing the detailed information which they felt they needed. The O and M Manager did not know the answers, the Computer Manager was too new to be able to provide them.

The arrival of the Computer Manager was followed by three weeks of discussions with Central Computer Services and the computer manufacturer in order to work out a policy for computer systems over the next five years. There were no plans to involve the head of O and M in these discussions, nor were departmental managers involved, although both groups had expected that this would happen. 'We'll call them in as and when they are needed' said the Head of Central Management Services.

In practice, each manager was brought in for discussions lasting around two hours and asked to outline the kind of assistance that he wanted from the computer. This was written down and agreed by the manager in question. It seemed that local groups, although given responsibility for the eventual success of the system, were being excluded from planning and policy making.

Positive communication reduces uncertainty

Some information was now being provided for those staff likely to be affected by the new system. A number of presentations were made to the stock control department by the Head of Central Management Services. These meetings were enjoyed and found to be useful. At them staff learnt the time scale of events and it was emphasised that the computer would bring benefits, not disadvantages. These sessions did a great deal to raise morale and to provide positive expectations of what the computer could do. Employees were still not able to get answers to their more detailed questions, however, as the system was not sufficiently finalised for these to be given. By mid-June morale was still quite good though it was now becoming clear that earlier time schedules would not be met. The original specification had stated that by June all computer staff would be engaged, the system would be designed and half the system specifications written.

Unclear task boundaries

It now seemed that people who could have been expected to play a major role in planning the system had not done so because they were unsure of where the boundaries lay between their task responsibilities and those of another group. For example, the work which the O and M Department had carried out was not sufficiently detailed for the preparation of specifications for the computer system. Similarly, line management had expected that Central Computer Services would tell them what to do and had done very little themselves because they had received no instructions. An early plan, set out in the feasibility study, had been to send all managers on a computer course. But by June no one had attended such a course and this seemed to be a result of the confusion over whether local or central groups had responsibility for communication and education. Nominally the responsibility lay with Central Computer Services, but the fact that this department was located in another part of the country meant that little formal training had, in fact, taken place.

In June 1969, the Board of the local firm was told that the Computer Department was unable to computerise the manual system in its present condition. Directors agreed with the computer specialists that a complete reorganisation of the manual system was necessary before a computer system could be effectively introduced. Because of these problems it was decided that the delivery of the computer should be delayed by six months and that the firm's General Manager should accept overall responsibility for the introduction of the computer system.

New attempts at role clarification

Later in the month a supplement to the original feasibility document was prepared which contained additional suggestions on how the system was to be implemented. It proposed a computer steering group representing both Central Computer Services and local senior management. Responsibility for the success of the subsystems was given to departmental managers. For example:

1 . . . implementation must be controlled to ensure feedback to all concerned. . . . This will enable progress to be monitored so that any necessary action can be taken both by local management and the management of Central Computer Services.
2 It is proposed that overall implementation of the project should be under the control of the computer steering group who would have a Board member as chairman and the Computer Department Manager as secretary. Other members of this group would be the Central Computer Services Manager, the Sales Director, the Works Manager, etc.
3 Each major subsystem would be developed and controlled by the appropriate departmental head. This should ensure that the computer system for each operating function within the company is developed and becomes operational in accordance with the requirements of the person responsible for that function.

Finally the supplement went on to say,

The computer analysts have joined the Central Computer Services Department so as to commence the preparation of detailed computer systems specifications. It has been agreed, however, that these analysts carry out detailed investigations into existing stock control procedures with a view to making changes in these procedures where this is desirable.

In this way a sharing out of the tasks was formally documented and the responsibilities of the various groups made clearer than they had been previously.

Role uncertainty and ambiguity

The first six months of the new computer application were therefore characterised by considerable uncertainty over which individuals and groups within the firm and within Central Computer Services should accept responsibility for different aspects of systems planning and design. This led to confusion and, on occasion, to an avoidance of responsibility for fear of treading on the toes of another group. In turn, this had the effect of slowing up the planning process. The firm at this time was characterised by 'role ambiguity': individuals and groups holding particular roles were uncertain about what was expected of them and others who had to interact with these individuals or groups were uncertain about what their expectations ought to be. There was a general lack of clarity about the roles associated with the introduction of the new computer system.

This lack of clarity may arise because of an organisation's inexperience with this kind of role and therefore an absence of clear expectations or it may be because the role incumbent is himself unsure of where the boundaries of his responsibilities begin and end. Both these conditions were found in our case study firm.

Uncertainty through absence of knowledge

The modified plan was now as follows: a steering group was to be set up to take overall responsibility for the installation with the Works Manager carrying ultimate responsibility for successful implementation. Project groups would organise the development of subsystems and these would be guided by the heads of the various functional departments. The broad terms of reference of the various groups remained as set out in the original feasibility study report with the systems analysts undertaking detailed investigations and making recommendations on future computer systems. Information on these various task divisions was put on the works notice board and sent to all line managers. The report given to line managers included the statement that the staff of the computer department would –

> investigate the present stock control methods over the next two or three months. After this investigation has been carried out line management will be consulted in order to establish the most desirable system. . . . It is important to realise that this assistance from Central Computer Services will only be available for the stock control exercise, and all departments must be responsible for future specifications and implementations of systems.

This statement placed responsibility for future systems fairly and squarely with line management, and it added to the anxiety and uncertainty of a management group which still did not understand what introducing a computer system really meant and also had serious doubts about its own competence to undertake this activity. Managers felt that they had responsibility but not the necessary knowledge to go with it.

Anxiety was increased by an interviewing programme carried out by the systems analysts, for many line managers found that they were unable to answer the systems analysts' questions about their work. Some now began to wonder if they had really understood the job for which they had been responsible over the years. One suggested that his activities were only a fraction of those he should have been carrying out. 'In fact all I was doing was progress chasing.' But he puts the blame for his erroneous definition of his duties on top management, saying that he should have been given some training when he was first moved into the post. 'How was I to know, I had to learn on the job?'

The new computer system was, in fact, now acting as a catalyst by making managers aware of wider aspects of their jobs which they had previously ignored or not been aware of. As a result, a loss of confidence set in among the managers and they attempted to lay the blame for what they now recognised as their deficiencies, on top management, saying that their jobs should have been defined more precisely. The reaction of top management to this was to suggest that line managers did not have enough initiative, they were not dynamic and had few ideas of their own.

The demand for a redefinition of role

The management group were now facing a dilemma common to most change situations. The abilities on which their previous success as managers had depended were becoming irrelevant and they were being expected to replace these with a new set of 'change' skills. Whereas previously they had been rewarded for carrying out a narrow range of duties in meticulous fashion, they were now being asked to take a critical look at their activities, to review their responsibilities and to take an imaginative and innovative approach to the problems underlying the work of their departments. They were expected to propose new methods of work based on the use of a computer. It is not surprising that they were anxious and uncertain about their ability to do this.

Our case study firm typifies the situation which is common when major technical changes are being introduced. At this time one set of expectations regarding a particular occupational role or set of roles is replaced by another, very different, set of expectations, and the role incumbents are required to be able to meet this new definition.

A feature of the introduction of the new computer system was that for the first time the firm was faced with having to deal with experts, highly qualified young men who asked searching questions and who could not be hoodwinked, yet who knew and cared little about the history and culture of the firm. Such experts, as we have seen in our other case studies, are easily stereotyped as an undesirable, alien group.

By September, the systems analysts had completed their investigation and a long report was presented to the Board. In effect they told the Board that, in their view, line management would have difficulty in assuming a new role. They said:

> Current systems are not capable of responding to demands of company expansion and computer introduction . . . the systems now operating must be critically reviewed . . . the major problem to be overcome is lack of systems discipline . . . this appears to have arisen from a lack of detailed procedure manuals . . . and rapid expansion which placed pressure on systems.

The report continued:

> Current management problems will also affect utilisation of computer services: the most effective computer systems are developed through a combination of line management experience and new ideas, together with computer and systems expertise of computer personnel. Line management is able in this way to identify with new systems and develop a vested interest in their successful implementation. . . . The existing management, however, to a large extent do not possess sufficient expertise in their own field to be able to identify what they require from a computer and would therefore find difficulty in utilising the information such a system produced.

This seems a perceptive analysis of the firm's dilemma – a dilemma shared by many other companies when introducing computer systems for the first time – that although there was a great willingness to take advantage of the new technology, there was not sufficient computer or systems expertise in line departments and plans had not been made soon enough to ensure that line departments were given this knowledge at an early stage of the innovation process.

How the Computer Department saw its role

Computer Department staff, though they had identified the local problem were not very sympathetic or anxious to take responsibility for its solution. They did not see themselves in a 'helper' role but would have liked the parent company to take executive responsibility for the change. Their attitude was 'if the user doesn't know what he wants, that's his problem'. Nonetheless the fact that the problem existed caused them a great deal of frustration because it hampered their work. The two systems analysts wondered why the parent company had agreed to the computer system when the local firm was so unready for it. Why also, they asked, did Central Computer Services not assume control and tell local management what it had to do. They were, perhaps, not as aware as their own manager of the political delicacy of this kind of situation. After all the firm had voluntarily requested that it should have a computer system.

The approach of the Central Computer Services Department in the early days of the installation had been as follows: users and O and M were to decide on an improved form of work system. This would then be presented to the Computer Department who would analyse it for computer application and suggest necessary modifications. But, as we have seen, this was asking too much of a local management inexperienced in both problem analysis and computer usage. Therefore, in December 1969, the computer staff redefined their own role and responsibilities and the order of work was reversed. It was decided by the Computer Manager that the new system must 'go live' on 1 July 1970, therefore the Computer Department must assume a more definite role than it had taken previously and play an active part in specifying the new system. As it had now been decided to use the computer manufacturer's stock control package, design flexibility was reduced and design procedure restricted by the constraints imposed by this package. The order of work therefore now became: the Computer Department decided on the system; they informed the O and M department, who gathered the relevant data from users. Any user requirements were now limited both by the July 1st deadline and by the dictates of the manufacturer's package. This placed the O and M Department in a buffer role between the computer specialists and the user departments. It left them with the difficult task of convincing the users that they would in the end get what they wanted from the new system.

Role clarification and the reduction of uncertainty

This more structured approach led to a clearer understanding of what was an appropriate division of responsibilities and to agreement on how different tasks should be allocated between the Computer and O and M Departments. The two groups came to a *modus vivendi*. The user managers also had their role clarified in a manner that gave them a greater sense of security although a reduced ability to participate in systems planning and design. But it can be argued with hindsight that Central Computer Services' original expectations of line management, although extremely democratic, were asking too much of a group used to routine kinds of work and experiencing their first computer installation. They therefore were a major source of uncertainty for this group.

By July 1970 the uncertainties of the previous phases of the project appeared to have been largely overcome. There was agreement on who should be doing what and on where the boundaries between the work of the different groups should be drawn. The expectations each group had of the others had become much clearer. The problem that now confronted them all was one of conflict rather than uncertainty. Would they all be able to play the roles assigned to them without serious disagreements breaking out? The users, in particular, were unhappy and antagonistic, because they felt that they had become powerless to influence what was happening to them. Much of their aggression was directed at O and M who were pressuring them for information, yet at the same time it was recognised that it was the July 1st deadline that was causing O and M to exert this pressure. User conflict during this period was therefore not so much with the O and M Department, whose problems they appreciated, but with top management and with the proposed computer system. Even at this late stage they were still uncertain of what the computer was going to do to them. Yet despite these new difficulties the computer system was implemented as planned on July 1st. This was the end of the implementation problems although operational problems were to continue until the inevitable bugs were ironed out of a new system.

Conclusion on case study

This case study illustrates another aspect of the uncertainty with which an organisation has to cope when introducing major innovation. This time uncertainty arose because of unclear role responsibilities and an unequal distribution of power and knowledge. The major difficulty was due to the fact that responsibility for different aspects of the planning, systems design and implementation processes had to be shared out between a number of interacting groups. Because the tasks associated with these activities were new to the firm there were no accepted and structured methods for allocating responsibility between the groups and there was a great deal of uncertainty on the part of each group with regard to what the other groups were or should be doing. In addition there were erroneous expectations of levels of competence, with the result that certain groups, in particular line management, were expected to assume responsibility for tasks which they had never tackled before and did not fully understand.

Our case study shows how local managers were uncertain regarding the role they should play during the installation period and how different managers had conflicting ideas on task responsibilities. For example, the O and M Manager defined his responsibilities in a very different way from his Board level superior. Again, there was great uncertainty over where the boundaries should be placed between the responsibilities of the Central Computer Services group and those of the O and M Department. In this situation the resources of power and knowledge lay with the external group and so the O and M group were unable to define their own role but had to accept that allotted to them by the staff of the parent company.

If either the local O and M or user groups had had greater knowledge of computer systems and of how to manage this form of change their bargaining position would have been stronger and they would have been better placed to specify the parameters of their own roles.

The O and M Department were uncertain of their responsibilities both to local management and to Central Computer Services. In the first relationship they had little guidance on the extent of their responsibility and authority and the O and M manager's ability to negotiate on this issue was hampered by the part-time nature of his assistants and their dual allegiance to O and M and to their own departments. In the second relationship the perceptions of O and M on what they should be doing did not fit in with the expectations of Central Computer Services. Central Computer Services, for its part, could not sustain its original intention of acting in an advisory and service role. This role specification had to be modified as computer staff recognised that their system objectives would not be achieved unless they accepted a greater degree of executive responsibility.

Thus all groups had particular problems in defining and assuming the required responsibility for the early introduction of the new system. User departments, in particular, had no experience of this kind of change and participating in it implied, to them, the acceptance of considerable risks. Their reaction fits with a statement by Rubenstein *et al.* (1967) that the knowledge of a new technology possessed by managers is a key factor in their ability to use it effectively. Similarly, the confusion which our managers showed agrees with observations by Mann and Neff (1961) who argue that, 'an individual's reaction to a change appears to be directly related to the clarity of his perceptions of the meaning of the change and his evaluation of the effect that the change will have on him as an individual with certain aspirations and expectations'. This absence of clarity regarding the content and boundaries of new roles is likely to produce uncertainty, an unwillingness to take decisions or responsibility, and may ultimately lead to attempts to withdraw from a stressful situation and leave someone else the responsibility for coping. All these behaviour patterns were discernible in our case study firm.

Organisations which are having to cope with uncertainty may seek to reduce this through trying to achieve a high degree of role consensus. By role consensus we mean the drawing of agreed task boundaries, so that the tasks for which one group is responsible are clearly delineated from those of another group. Role consensus may be achieved in a number of different ways. It may come about through *standardised procedures* (Thompson, 1967). This will happen when job analyses are carried out by the personnel department and job specifications drawn up and placed in the files of the personnel department. This is to some extent an imposed consensus with an external agency, the personnel department, specifying the job and requesting agreement with its definition. Role consensus may also come about through *planning*. With this approach problems are analysed strategically and responsibility for the different tasks associated with their solution allocated amongst different groups. This can also be an imposed consensus, perhaps by top management, but it can also be a recognition by the interacting groups of where different competences lie.

Lastly, role consensus may, as in our case study firm, be a result of a process of mutual adjustment. This involves the testing out of competences in an action situation, and effective communication and feedback between the groups until task boundaries are drawn and accepted. In the firm which we have examined agreement on the allocation of tasks was reached but not necessarily liked and user departments believed that they were giving too much power to the computer specialists and losing the ability to control their

own situations. Nevertheless there was eventual agreement on role definition.

Some organisations may respond to uncertainty in a different way by keeping roles fluid and bringing people together in problem-solving groups when
the need arises. Role definitions are then loose and temporary and related to
needs at a particular moment in time. This kind of matrix organisation is still
unusual but is starting to be used as a means for assisting innovation. It has
the advantage of being highly responsive to change in the firm's or department's environment but it has its own built in element of uncertainty for
there is evidence that it leads to effective problem-solving but also to stressful
interpersonal relationships.

8

Political aspects of the specialist-user relationship

In earlier chapters of this book we argued that innovation is often a response to environmental change and pressure. Faced with some alteration in their product market or technological environment firms respond fairly predictably by creating a new department or subsector of the business to manage this new environmental input. Thus we have new venture study groups to scan the environment for potential growth areas, and mergers and management services departments and organisational development departments to introduce the technologies of computing and organisational design and change. But the problems of new organisational units lie not so much in their act of creation, though this may also be a minor drama in itself, but in the management of their integration into the organisation over time.

New organisational forms create additional uncertainties. Where is the unit to be located in the organisation's structure? What is its distinctive competence to be? Who should lead and staff it, whose needs can it satisfy? These questions, while easy enough to state, are considerably more difficult to disentangle in any empirical context. Just as innovative decision processes are often characterised by internal politicking, so the initiation and evolution of new organisational units are often bounded by political forces within the organisation.

The new specialist department either by its structural location or reporting arrangements, by the way its staff are differentially rewarded or through the manner in which its tasks are linked to the territories and personal ambitions of others, can affect the prevailing distribution of power and status in the organisation. It is chiefly the uncertainties about existing and future allocations of activities, rewards and associated power and status systems both between specialist and user and new specialist unit and existing specialist unit, which make the integration of the new unit a continuing source of organisational tension.

For the individual specialist, however, strategic questions about the overall integration of his unit into the organisation may be less real than the specific political constraints and opportunities afforded by the project he is currently working on. He may see this project as an opportunity to develop a particular technical process, or a chance to redefine another sector of the business within the logic of his own approach or a chance to work with a new powerful user group in the business. To the latter the specialist's intrusion may appear as a threat to his well ordered, maybe self-formulated way of tackling things. It may also represent a specific encroachment on his present status or future career opportunities. The choice of project area, the composition of the project team, the way the assignment develops and its likely acceptance and implementation by the user are all important transactions bounded by political forces within the organisation. Whether the specialist appears in these transactions as part of a ginger group, a helper, a catalyst or an expert technician, what is taking place in such transactions is a mutual influence process between specialist and user. This chapter presents one viewpoint on the specialist's role in that influence process.

Internal politics and organisational change

Political activity in organisations tends to be particularly associated with change (Pettigrew, 1973). Since specialists are the initiators of many organisational changes their activities and plans are inextricably bound up with the politics of change. Major structural changes, or even the possibilities of them, have political consequences. Innovations are likely to threaten existing parts of the working community. New resources may be created and appear to fall within the jurisdiction of a department or individual which had previously not been a claimant in a particular area. This department or its principal representative may see this as an opportunity to increase its power, status and rewards in the organisation. Others may see their interests threatened by the change, and needs for security or the maintenance of power may provide the impetus for resistance. In all these ways new political action is released and ultimately the existing distribution of power is endangered.

The above analysis has suggested that the specialist–user relationship takes place in the context of organisation life where political activity is pervasive and real. Furthermore, the activities of specialists, especially in so far as they demand structural changes in organisations, will affect the current balance in the distribution of power and thereby involve both user and specialist in those political processes. If that involvement is not proactive, then it will be reactive as the political behaviour of others acts as a constraint on the range of behaviour possible for both user and specialist. In this situation, as Bennis (1969) has suggested, if the adviser bases his approach on two sources of influence, truth and love, then it seems likely his plans will remain only dreams. While strategies of influence based on organisation development norms may be 'appropriate under conditions of truth, trust, love and collaboration' they may be much less appropriate under the political settings described elsewhere by this and other authors (Dalton, 1959; Crozier, 1964;

Pettigrew, 1973a). Part of the 'valid and useful information' not covered in Argyris's (1970) theory, yet required by the specialist, is a knowledge of the political processes in his own organisation and an awareness of how the particular projects he is working on relate to, and by implication, alter, those processes.

An additional contextual factor which can greatly constrain the specialist's interpretation of his role is the stress to which he is exposed. In setting out a rather different theory of intervention from the present one, Argyris emphasises the importance for the consultant's effectiveness of his ability to accurately perceive stressful reality. This would seem to be a critical yet relatively unexplored issue.

Stress in the specialist's role

Argyris (1970) has described with great insight some of the stresses and defences in the user–specialist relationship.* His analysis is particularly strong on some of the interpersonal and psychological causes and consequences of ineffective intervention activity. Once the specialist finds it difficult to perceive stressful reality accurately, to value user attack, to trust his own experience, and to invest the stressful environment with growth experiences, the following consequences are likely. First, increased specialist defensiveness with a corresponding need for competence and compulsiveness for success. Secondly, decrease in the use of appropriate mechanisms with a corresponding increase in psychological tiredness and decreased tolerance for stress and ambiguity. As the specialist's anxiety about his lack of success increases so his need for inclusion with and confirmation from the user increases.

The present analysis seeks to complement Argyris's approach in three ways. First by raising some of the *structural* sources of stress in the specialist's role; secondly by raising some additional interpersonal consequences of this stress not fully explored by Argyris; and finally, by discussing the impact of the above two factors not only on the direct relationship between user and specialist but also on the relationship between specialist and specialist. Suggestions are then made about how processes going on between specialists can affect the power relationship between specialist and user.

The meaning of stress

Stress, like job satisfaction, is an umbrella concept which is made up of a number of components. It is usually defined as a response based phenomenon. This means that the existence of stress is inferred from an individual's responses in a particular situation or set of situations. Thus a specialist may be in a stressful situation if he is responding to his environment by being

* He presents his argument in the language of client-consultant and focuses particularly on Behavioural Science Consultants, but there are many parallels between his consultants and our technical specialists.

aggressive, impulsive and rigid. The stimulus for the stress is usually talked about as an environmental demand.

The most convenient way to examine the causes and consequences of stress is to look at it as a four-phase sequence.

Figure 8.1 recognises that a critical part of the stress process is the perception by the specialist of the environmental demand. An important complicating factor for all discussions of stress, whether they relate to causes or consequences, is the phenomenon of individual differences. Different individuals perceive the same objective conditions in different ways. This means they are likely to experience stress and respond to it rather differently (Pettigrew, 1972a). Nevertheless, enough is known about stress to indicate that certain structural components of roles tend to lead to the generalised experience of stress. These *job stressors* are frequently discussed as role overload, role underload, role conflict and role ambiguity (Kahn and Quinn, 1968; Kahn *et al.*, 1964; Pettigrew, 1968).

Fig. 8.1 Phases and components of stress

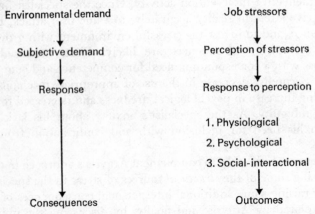

A generalised model for describing the role overload–underload situation is presented in Fig. 8.2. This model suggests that stress occurs when there is a substantial imbalance between a perceived environmental demand and the response capability of the focal person. It is assumed that stress will only be felt when the consequences of failure to meet the demand are important.

Fig. 8.2 The overload–underload stress condition

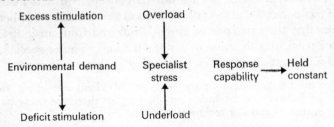

Figure 8.2 indicates that with the response capability of the focal person held constant, overload results from an inordinate increase in environmental demands, while underload results from the reverse process. Stress is most obvious in roles where the individual has little control over the volume of work coming to him, where there are large peaks and valleys in the flow of work and where such fluctuations are not predictable. Studies of specialists in the Management Services field by Mumford (1967) and Pettigrew (1973b) suggest tentatively that role overload and underload and the combination of the two are to be found in systems analyst and programmer roles.

In the Mumford study, 73 per cent of the systems analysts and 81 per cent of the programmers acknowledged that their work could be stressful. The following quotations give some indication of the overload–underload problem:

- The speed at which we sometimes work is stressful.
- Communications between us and management are often deficient. Occasionally they want to know why this job wasn't done yesterday.
- I have a dozen things to do at the same time. Yet I prefer to do one thing and then move onto something else.
- If you've got a deadline and you can't meet this, people start pushing you and the whole thing becomes a nightmare.

The other two principal job stressors are role conflict and role ambiguity. Role conflict occurs when a person is subject to incompatible role expectations. There are a number of different types of role conflict. We are especially interested in person-role conflict. Kahn *et al.* (1964) define this as 'the conflict which may exist between the needs and values of a person and the demands of his role set'. Role ambiguity refers to the inadequate sending of expectations and obligations to the focal person from the members of his role set.

Evidence from a study of another group of specialists, operations researchers, gave strong indications of the existence of person-role conflict and role ambiguity. In that study it was found that the OR man's role orientation lay towards satisfying his desire for creativity, autonomy and variety. He wanted to satisfy the *research* side of his job title. Meanwhile the organisation was pressuring the OR man towards the *operational* side of his job, which affects the time he can take over projects, the nature of the projects themselves (in this case they are too routine) and the reception which his findings are given (Pettigrew, 1968).

The sources of role ambiguity for the operations researcher were the terms of reference and information given by clients, company general policy and its relationship to the projects they worked on and the scope of their own responsibilities and authority.

- You cannot be sure how far your responsibility is allowed to go. I'm not clear in a formal sense what I can and cannot do in a particular job.
- I am not clear – there are points which might benefit from clarification. If

I clarified my position too much, people might think I was usurping their spheres of influence.

In spite of the lack of research in this field, there is evidence to suggest that the four job stressors of overload, underload, conflict and ambiguity are built into certain specialist's roles. The issue is, what *consequences* does this stress have on the specialist's ability to influence user departments? A look at some of the coping responses specialists make under stress might suggest that some specialists are not fully aware of the range of power resources available to them and how they might mobilise and tactically use power in relationships with user groups.

Some consequences of stress for the specialist's power

Before discussing the sources and use of specialist power it is important to acknowledge the impact of stress in the specialist's role on his ability to influence user groups. The literature on coping responses under stress is at the moment fragmentary and incomplete. There are certainly few, if any, comprehensive studies of the coping responses of specialists under stress. In this situation, one can hypothesise on the basis of generalisations from laboratory studies (Lazarus, 1966) and the few field studies that exist (Kahn et al., 1964; Appley and Trumbull, 1967; McGrath, 1970).

Perhaps the best-documented responses to stress are those that impair the individual's cognitive and perceptual ability. These have been most comprehensively discussed by Kahn and Quinn (1968) and Deutsch (1969). Kahn and Quinn mention increased rigidity, aggressiveness, selective perception and intolerance for ambiguity. Deutsch presents his argument within the framework of what he calls the destructive course of conflict. He lists, in effect, some of the well-known indicators of cognitive collapse:

1 a reduced range of perception: fewer alternatives are seen;
2 a reduced time perspective: the individual focuses on the immediate, rather than long-term consequences of alternatives;
3 increased rigidity and polarisation of thought processes;
4 increased impulsiveness and defensiveness.

Both Kahn *et al.* and Deutsch suggest that the above set of responses to stress are likely to compound the pre-existing interpersonal difficulties in any relationship. As the original job stressors of overload, underload, conflict and ambiguity tend to have secondary effects, so the processes of cognitive collapse feed back into the original interpersonal problem. As we shall see later, the misjudgments and misperceptions arising from the above responses to stress have a major impact on the processes of impression formation between specialist and user.

One very specific response that certain individuals make under stress is to withdraw from the source of stress (Lazarus, 1966). This is perhaps one of the most common responses made by specialists. The specialist either with-draws into himself and becomes so preoccupied with the intricacies of his own expertise that this becomes his only way of attempting to influence the user, or he regresses interpersonally and only sees the user when a specific task issue is at hand. Both responses produce secondary effects which are unlikely to lead to successful specialist influence attempts.

The sole reliance on expert power through the demonstration of technical competence is rarely successful against a user probably already very defensive about his own lack of technical expertise. Baker and Schaffer (1969) argue that inadequate user–specialist relations are often made worse 'by the be-haviour of the staff consultants, who mask their own unsureness and anxiety with a thick amount of professional jargon and technical talk'.

Interpersonal regression is a further common specialist response under stress. The present theory of social power argues that the power aspirant, in this case a specialist, is much more likely to influence the user if he has established a *multiplex* relationship with that user than if he had formed only a *uniplex* relationship with him. The notions of multiplex and uniplex rela-tionships are taken from theories of social anthropology. Max Gluckman (1956) used the notion of multiplexity. Kapferer (1969) has perceptively developed their analytical use in his research on social networks.

All interactions are composed of exchange contents. These are the overt elements of the transactions that take place between the individuals in the interaction. They might be conversation, personal service, job assistance, cash assistance and task-related interactions as distinct from social related interactions. The latter, in an organisational setting, might be spending coffee or lunch periods together and a range of after-work social and sporting contact. Multiplexity refers to the number of exchange contents in any rela-tionship. A relationship becomes multiplex when there is more than one exchange content within it. A uniplex relationship has only one exchange content. It is assumed that multiplex relationships are 'stronger' than those which are uniplex. Generally speaking the specialist will be able to exert greater pull and influence over the user to whom he is multiplexly tied.

A major, and as yet untested, hypothesis is that many specialists under stress conditions regress interpersonally. Some specialists never get beyond an exclusively task-based relationship with users. They only ever form uni-plex relationships. The present suggestion is that those who are able to build multiplex relationships find them very difficult to maintain over time. Under stress they are likely to regress to the uniplex relationship where their pre-occupation with task and their own expertise offers a temporary, if deceptive, haven of stability and security.

The sources and use of specialist power

Earlier it was argued that existing attempts to describe and conceptualise the user–specialist relationship have been found wanting. Bennis (1969) addresses himself in general terms to the problem: 'Putting it differently, there seems to be a fundamental deficiency in models of change associated with organisation development. It systematically avoids the problem of power, or the politics of change.'

The present analysis seeks in part to deal with that deficiency in the literature. First, by recognising the essentially political character and consequences of organisations undergoing change. Secondly, by acknowledging that most specialists must interpret their roles in the context of their own proactive involvement in those political processes or react to the constraints imposed on them by the political activities of others. And finally, by offering a way of conceptualising the political behaviour of specialists by focusing on the potential power resources they possess.

Power is not an attribute possessed by someone in isolation. It is a relational phenomenon. Power is generated, maintained and lost in the context of relationships with others. A power relation is a causal relation between the preferences of an actor regarding an outcome and the outcome itself. Power involves the ability of an actor to produce outcomes consonant with his perceived interests. Most actors have power only in certain domains of activity. The scope of their power is limited by their structural position in their organisation. This is because the resources which form the base of an actor's power are differentially located by structural position. In this sense, the transferability of power across system boundaries is regarded as problematic.

Power resources must not only be possessed by an actor, they must also be controlled by him. Bannester (1969) makes this point succinctly: 'It is immaterial who owns the gun and is licensed to carry it; the question is, who has his finger on the trigger?' Control, however, may not be enough; there is also the issue of the skilful *use* of resources. The successful use of power is also a tactical problem. The most effective strategy may not always be to pull the trigger.

From this viewpoint the analysis of organisational power requires some attempt to map out the distribution and use of resources and the ability of actors to produce outcomes consonant with their perceived interests. As we shall see, the main practical problem for the specialist is the movement through the stages of possessing, controlling and tactically exploiting the power resources he possesses. Some specialists are not aware of the potential power resources they do possess. Others are aware of the resources but can neither effectively control or tactically use them. But what are some of these potential specialist power resources and how and why might they be used?

Pettigrew's (1968, 1973a, b) research on systems analysts, programmers and operations researchers indicated that there are at least five potential power resources available to these internal consultants. These are:

1 Expertise.
2 Control over information.
3 Political access and sensitivity.

4 Assessed stature.
5 Group support.

These resources are separated here for analytical clarity; they are, of course, empirically highly interdependent.

Expertise

Singular possession of a valued area of technical competence is perhaps the most familiar source of specialist power. In one study, 70 per cent of a sample of operations researchers felt they had an influence on company decision-making. Asked how they thought they were able to exert this, 55 per cent said 'because we alone have the time and techniques to produce detailed and novel solutions to complex problems of planning' (Pettigrew, 1968).

The notion of dependency is crucial to the analysis of expert power. Emerson (1962) supplies the initial exploration of dependency. Blau (1964) interprets him as follows: 'By supplying services in demand to others, a person establishes power over them. If he regularly renders needed services they cannot readily obtain elsewhere, others become dependent on and obligated to him for these services', unless they in turn can supply services which he also needs. The power of one individual over another thus depends on the social alternatives or lack of them available to the subjected individual.

The power of specialists over users is likely to be consequent on the amount of dependency in the relationship. The specialist can maintain a power position over users as long as they are dependent on him for special skills and access to certain kinds of information. One way that specialists can generate such dependency is to manipulate the uncertainty surrounding their expertise (Crozier, 1964; Pettigrew, 1973b).

The power of the specialist is unlikely to be omnipotent even with the most technically uncertain problem. Clearly most dependency relationships will be a matter of degree. The relative *centrality* and *substitutability* of the specialist group is likely to vary over time (Hickson *et al.*, 1971). In times of financial stringency the accountant's activities may become more central to the organisation's survival. In times of the proliferation of specialist groups all operating from a similar task environment, one group's activities may be seen by user groups to be readily substitutable for another. (One strategy used by users to weaken specialist power is to create another source of specialist expertise and encourage competition between the two sources. External consultants may be brought in for this purpose [Pettigrew, 1973b].) In both these situations dependency relationships may be difficult to generate and maintain.

Control over information

Several authors have mentioned the control over information as a power resource. Mechanic (1962) argues that within organisations dependency can be generated for others by controlling access to the resources of information, persons and instrumentalities. Burns and Stalker (1961) assert that information may become an instrument for advancing, attacking, or defending status. Using the prison as a setting, McCleery (1960) is able to demonstrate how the formal system of authority relations may be considerably modified by the location and control of communication channels. Because all reports had to pass through the custodial hierarchy this group was able to subvert the industrial and reform goals represented by the Prison Professional Services and Industry programmes. The head of the custodial hierarchy, the prison captain, for the same reasons was able to exert considerable control over decisions made by his immediate superior, the warden.

The structural location of many internal specialists offers them particular advantage with regard to the control over organisational communications. Most specialists have roles with high boundary relevance. They have many significant work contacts across departmental boundaries within their own organisation and between that organisation and relevant others. In this regard, they are well positioned to take on the role of technical gatekeepers. As such they are potentially able to influence the resource allocation process in their organisation through a process of collecting, filtering and reformulating information (Pettigrew 1972b).

Research by Pettigrew (1973b) has indicated consultant gatekeepers may be particularly effective at controlling information in the uncertain conditions surrounding innovative decisions. During these decisions strategies of uncertainty absorption may enable specialists to produce outcomes in line with their perceived interests (March and Simon, 1958). Counterbiasing by user groups in these decisions is likely to be difficult. In the specialist–user relationship where the information passed is likely to be complex, uncertain and rapidly changing, the possibility for user groups either to identify bias, or to deal with it by counterbiasing, is likely to be more difficult.

Aside from their ability as gatekeepers to control the flow of technical information, specialists may have access to other sorts of information which they may be able to put to use. As part of their investigations into other departments, specialists may uncover the inefficiencies and incompetences of others. Even if they do not actually reveal these inefficiencies, users may *perceive* that the specialist has this information. Some specialists are prepared to use this information against recalcitrant clients. It is frequently referred to as the 'dirty linen strategy'.

Few specialists can afford to rely exclusively on their expertise and control of information. The specialist's placement in the communications structure needs to be reinforced by other forms of political access. The specialist may not just rely on the presumed dependency which the mystique of his expertise can give him. He may actively seek support for the demand he is making. His ability to generate this support will be conditional on his possession, control and tactical ability in using three additional power resources. His assessed stature in the locus of power in his organisation. The amount of political access he has and its corollary, his sensitivity and use of political information and processes. And finally, the amount and quality of group support he can mobilise from his peers. As we suggested earlier, the ability of specialists to influence user departments is a function of at least two interdependencies. That between specialist and specialist and that between specialist and potential user. Power is a systemic and not just a relational phenomenon.

Political access and sensitivity

To the specialist interested in the acceptance and implementation of his ideas, political access is likely to be critical. In the political processes which surround many organisational changes the ideas which remain supreme will not necessarily be a product of the greater worthiness or weight of issues ranged behind them, but rather in the nature of the linkages which opposing parties have to individuals over whom they are competing for support. The amount of support a specialist achieves is likely to be conditional on the structure and nature of his direct and indirect interpersonal relationships.

The present discussion of political strategies is based on the assumption that men seek to adjust social conditions to achieve their ends. This view of man does not assume that *all* behaviour is self-interested. Neither is it assumed that the process of adjusting means to ends is a rational one. Individual choices are limited by their perception of the situation in which they act and by the amount and accuracy of the information on which they base their strategies. As Kapferer (1969) notes, individuals continually commit errors because of misperception through lack of information or miscalculation. They may also be manoeuvred into committing error. A final and significant restriction on rationality is the constraint on access imposed by man's location in a network of social relations.

The politically aware specialist is likely to be conscious of the span of his social network, the degree of reciprocity in these contacts and the extent to which the relationships are uniplex or multiplex. How focused the specialist's network is to the locus of power in the organisation is also critical. Clearly there is little political advantage in having a network with an extensive span of multiplex relationships if they are with individuals with little power. The specialist should be sensitive to the relative power of those he endeavours to attract. Along with a reasoned perception must come effective action. Bailey (1969) makes this point well:

> Knowledge is power. The man who correctly understands how a particular structure works can prevent it from working or make it work differently with much less effort than a man who does not know these things. This may seem obvious yet actions are often taken without previous analysis, and out of ignorance!

Assessed stature

It should be clear from the preceding discussion that the mobilisation and tactical use of power rests on the understanding of at least three general elements. A power aspirant with his set of potential power resources. Some mode of interpersonal activity or other communication, and a recipient of the influence attempt. The notion of assessed stature not only seems to be a critical power resource for the specialist; it also serves as a way of conceptually linking the above three elements of the power process.

Specialists do not merely advise, they persuade, negotiate and exercise the power they can mobilise. Assuming the specialist is both able to identify successfully, and has access to, the centre of power in his organisation, an important constraint on his ability to negotiate and persuade is likely to be his assessed stature, both in that centre of power and in his immediate interpersonal relationships with clients. *Assessed stature is defined as the process of developing positive feelings in the perceptions of relevant others.*

The development of assessed stature may be likened to what Goffman (1969) has to say about impression management:

> When an individual enters the presence of others, he will want to discover the facts of the situation. Were he to possess this information, he could know, and make allowances for, what will come to happen and he could give the others present as much of their due as is consistent with his enlightened self-interest.

What is critical, yet imprecisely defined by Goffman here, are 'the facts of the situation'. From the authors' viewpoint the process of accruing the facts of the situation may be conceptualised as identifying user department *role saliences.*

Before discussing the meaning of role saliences it may be opportune to quote some examples of what occurs when role saliences are not exposed by a party making an influence attempt. Studies by Triandis (1960a, b) on superior–subordinate relationships supported the general proposition that the greater the cognitive dissimilarity between two persons the less effective will be the communication between them and the less satisfied will they be with their relationship.

In a later study, this time of interpersonal relationships in international organisations, Triandis (1967) substantially upheld his hypothesis that work associates who belong to different cultures will experience severe communication problems and low levels of affect towards each other. A less precisely researched but more poignant example of the failure to anticipate role saliences in others is offered by Edward T. Hall (1968).

> Despite a host of favourable auspices an American mission in Greece was having great difficulty working out an agreement with Greek officials. Efforts to negotiate met resistance and suspicion on the part of the Greeks . . . upon later examination of this exasperating situation two unsuspected reasons were found for the stalemate: First, Americans pride themselves on being outspoken and forthright. These qualities are regarded as a liability by the Greeks. They are taken to indicate a lack of finesse which the Greeks deplore. . . . Second, when the Americans arranged meetings with the Greeks they tried to limit the length of meetings and to reach agree-

ment on general principles first, delegating the drafting of details to sub-committees. The Greeks regarded this practice as a device to pull the wool over their eyes. . . . The result of this misunderstanding was a series of un-productive meetings with each side deploring the other's behaviour.

Clearly what is established from the Triandis (1967) study and Hall's example of interpersonal misunderstandings in diplomatic negotiations, is that if parties to such relationships are to influence one another they must be in the position to identify what is *salient* in the other party's perspective and behaviour. The concept of role salience recognises that different groups of users and specialists have varying sets of needs, expectations and reference group affiliations. They also relate to others with differing sets of political interests. The present suggestion is that a specialist's ability to *anticipate* what is salient for the user in these terms is an important component of the process of generating a high assessed stature for himself. Clearly the assessment and anticipation of these saliences will be easier for a specialist with a social net-work with an extensive span, and with multiplex rather than uniplex relation-ships across that network.

In this discussion the concept of assessed stature is not equivalent to the French and Raven (1966) notion of referent power. The specialist is not trying to 'identify' or develop a 'feeling of oneness' with the user. Rather the specialist is seeking to identify and thereby anticipate what is salient for the user both in task and political terms, so that his proposals may be formulated to receive minimal user resistance and maximal support from the locus of power in his organisation.

In the early stages of any user–specialist relationship part of the tactics of generating high assessed stature may involve demonstrating competence in areas salient to the user. This has been described by some specialists as 'the low key approach'. The specialist takes on small jobs perceived to be salient for user needs and whose successful outcome can be priced in pounds sterling. In this way the specialist builds up credits for himself with significant others, and later is able to generate support for projects whose salience to others is not so readily discernible.

Alongside the specialist's ability to generate high assessed stature must come the ability to perceive when he has high and low stature. Power derived from stature is a variable phenomenon. Political timing, therefore, becomes important. The time and the way a proposal is presented may have a crucial impact on the support it receives. The specialist seeking to mobilise power must be careful to make his assertions at a time when he has the resources to enforce his will. The specialist with low stature does not make demands on the system that threatens him.

Group support

Arguments in previous sections have indicated that expertise, control over information, and political access and sensitivity are *necessary* but not *sufficient* conditions for specialist power. The possession and tactical use of these power resources needs to be considered in the context of the specialist's assessed stature in the social arena in which he works.

There is at least one other important variable feeding into this system, the amount and kind of group support given to the specialist by his colleagues in his own department and in related specialist groups. A major constraint on political activity in all organisations is the amount of time and energy so consumed. Research has indicated that protracted power struggles between specialist groups use up a great deal of the reserves of time and energy these groups might have used more profitably assisting one another in generating support for their ideas with user groups (Pettigrew, 1973b).

The coordination problems posed by specialist groups such as operations researchers, systems analysts and programmers relate particularly to their emergent status (Pettigrew, 1973a). The task environment shared by developing specialities is often poorly institutionalised. That is to say, the system of role relationships, norms and sanctions which regulate access to different positions and sets of activities lack both clarity and consistency. In the absence of such clarity and consistency a process of role crystallisation takes place. Strauss *et al.* (1964) have described the strategic aspects of this crystallisation as the negotiation of order.

Problems of status and power arise as the emergent specialist groups seek social accreditation. Some groups take on expansionist policies, intrude on others' domains and provoke conflict. The process of the conflict between the rival groups may take on the form of a set of boundary-testing activities. As one group seeks power and the other survival, each will develop a set of stereotypes and misconceptions about the other. A group declining in status and power may seek to emphasise that part of the core of its expertise which still remains and which may not be covered by the activities of the expanding group. This may be interpreted as a threat by the newer group who are likely to be defensive about their own history of inexpertise in this area. They in turn may retaliate by emphasising their particular strength. In this way one group's defensive behaviour becomes another group's threat, and the cycle of conflict continues.

The resolution of such structural conflicts seems questionable. The conflicts have been *regulated*, however, through a variety of integrative mechanisms. These include creating integrative roles, project teams, project controllers and training in interpersonal skills.

An additional problem remains. Apparently some executives are prepared to encourage conflict between and among specialist groups as a way of controlling them more effectively. Wilensky (1967) argues Roosevelt's technique for controlling his technical subordinates was to 'keep grants of authority incomplete, jurisdictions uncertain, charters overlapping'. Pettigrew (1973b) quotes the example of a Board of Directors' use of this approach.

By keeping distant from the scene of conflict, by giving the programmers some freedom from the system of bureaucratic rules, and by keeping job assignments uncertain, subject to change at any moment, they prevented the programmers from consolidating in a stable power base, and still managed to extract the knowledge and work necessary for the company's continued prosperity.

Power conflicts between specialists are likely to be a continuing feature of organisational life. Issues concerned with the relative share of interdependent activities and the distribution of status and power are only fundamentally defused when the groups are either no longer interdependent or the supremacy of one group becomes so clear that further protest from the other is perceived to be futile. In the continuing absence of either of these conditions the political position of specialists *vis-à-vis* user groups will be proportionately weaker.

Conclusion

This chapter has sought to emphasise the essentially political character of organisational life. Specialist–user relationships take place in the context of organisations where struggles for the advancement and maintenance of power and status are pervasive and real. The origins and momentum for these political processes are in part the continual changes that most organisations experience. Major structural changes have political consequences.

Specialists have a vested interest in change. This is how they legitimate their presence. Many user groups have a vested interest in relative stability. They have quotas to reach, deadlines to meet and empires to protect. There is no reason to expect them readily to accept changes which are against their interests. That is why the relationship between specialists and executives is regarded as a problem one.

This chapter has focused on the conceptualisation of the influence process between specialist and executive. In our viewpoint the influence process has to be understood in the light of two major contextual factors. First, the stressors of overload, underload, ambiguity and conflict built into many specialists' roles and secondly, the political character of much of organisational life. Specialists often experience considerable job-related stress. They are also aware that the process of generating support for their ideas is very much constrained by the political implications of their proposals. Equally well they are aware that under many circumstances their success is dependent upon their ability to mobilise power for themselves in the face of those constraints.

The present concern has been to conceptualise the mobilisation of specialist power. Power has been defined as a causal relation between the preferences of an actor regarding an outcome and the outcome itself. Power involves the ability of a specialist to produce outcomes consonant with his perceived interests. The base of the specialist's power rests on his possession, control and tactical use of five resources. These are expertise, control of information, political access and sensitivity, assessed stature and peer and related specialist group support. Of these the first three appear to be necessary but not sufficient conditions for specialist power. Once he has the political access and understanding, the specialist's ability to negotiate and persuade depends on his assessed stature with the appropriate figures in his political network.

Part of the process of generating high assessed stature rests on the specialist's ability to manage the impressions he has created with others. Given the relatively high levels of stress in the specialist's role and what we know about cognitive and perceptual impairment under stress, this process of impression management is not an easy one. It is suggested that under these conditions some specialists withdraw into their expertise, others regress interpersonally. Both these reactions to stress produce secondary effects which impair the specialist's ability to generate support for his proposals.

An important part of gaining high assessed stature appears to be the specialist's ability to identify and thereby anticipate role saliencies of potential users and/or key figures in his political network. These role saliencies refer not only to the user's values, expectancies and reference groups but also to his political and career interests and how they might be affected by any specialist proposals. Credit can be built up by attending to projects which relate particularly to user saliencies. This is especially so if the benefits of the project can be expressed unequivocally in financial terms. These credits may then be 'cashed' on projects the specialist has a particular interest in.

Finally, this analysis has highlighted the significance to the specialist of his ability to form multiplex relationships with key figures in his political network. On the development of these relationships depends the specialist's capacity to identify role saliencies in others and also to assess clearly and accurately when his stature is high and low.

9

Uncertainty and strategic decision-making

Uncertainty has been described by Lawrence and Lorsch (1967) as situations characterised by an absence of specific, relevant and timely information; situations with a long timespan of definitive feedback, where there may be a wait of months or years before the correctness of decisions and actions shows up; situations in which there is a lack of knowledge concerning cause and effect, in which people cannot perceive with any clarity the consequences of the decisions and actions which they are taking. The theme of this book has been organisational uncertainty when implementing decisions. We have tried to show that organisations, in an attempt to cope with uncertainties located in their external product or labour markets, introduce new ideas and techniques which they hope will enable them to gain a degree of control over these external environments. Increasingly the form of innovation selected is based on the use of computers. This technology is perceived as enabling the organisation to become more efficient by providing faster and more relevant information, thus enabling the firm to interact more successfully with its external environment and to meet more effectively the objectives which it is in business to achieve.

Large-scale computer systems have particular uncertainties associated with them. The design phase frequently takes place in an environment which is technologically dynamic with new and better methods for solving the problem continually appearing as a consequence of developments in hardware, software and peripheral equipment. Planning and implementation have to embrace a large variety of technical, social and economic factors, many of which may be overlooked until they are forced to management's attention by a danger signal of some kind. In addition the dynamics of this form of innovation and its non-programmed nature lead to political behaviour as groups and individuals seek to use the fluidity of the change situation to promote or protect their own interests. User departments, in particular, experience great uncertainty when faced with new computer systems; frequently they have little experience of either computer technology or large-scale change. Coping with it may bring interpersonal stresses which arise from the need to interact with new groups or groups which they normally prefer to keep at a distance. They will have to establish relationships with technical experts – the computer specialists – and they are likely to be subject to considerable pressure from top management for fast and successful results.

Figure 9.1 shows how in responding to external uncertainty with innovation the organisation generates considerable internal uncertainty. The very fact of change introduces an element of instability as long established methods and procedures are abandoned and new ones introduced which are better able to take advantage of the opportunities which computer technology presents. In this book we have attempted to set out the nature of this internal uncertainty and how it is dealt with. We have shown that the action of coping with certain kinds of uncertainty generates other kinds of uncertainty which in turn have to be dealt with. In this way the organisation gets caught in an interactive process of recognising uncertainty, coping with it and producing new uncertainty which only ceases when the innovation is successfully installed and becomes operational.

The second and third levels of the diagram illustrate the fact that the common group and individual response to uncertainty is an attempt to reduce it by gaining understanding of, and control over, the factors which are producing it. At the group level making plans and making decisions is one way of doing this and in chapters 2 and 3 we examined the extent to which planning principles and decision theory take account of the concept of uncertainty. But, like the organisation as a whole, groups and individuals in their efforts to reduce uncertainty generate new areas of uncertainty. Each major decision tends to produce a dependent set of sub-decisions of a technical or administrative kind. Similarly, the making of a decision requires a reconciliation of different interests and the process of reconciliation may show up new areas of disagreement and conflict.

Uncertainty and the organisation

A difficulty for the organisation which chooses innovation as a solution to its problems is that, like all organisations today, it has to cope with a dynamic external environment. Computer innovation is slow and it may take several years before a large system is implemented and becomes operational. During this period the firm may experience major changes in its technical, product and labour markets and decisions taken at an early state of a systems development may be no longer appropriate at a later stage. This environmental turbulence can mean that an organisation which uses rigid planning procedures runs the risk of ending up with an inefficient and inappropriate solution to its problems. We have suggested that decision-making needs to be associated with excellent monitoring and feedback mechanisms so that change in the environment can be quickly identified and decisions modified to take account of this.

Fig. 9.1 Organisational uncertainty

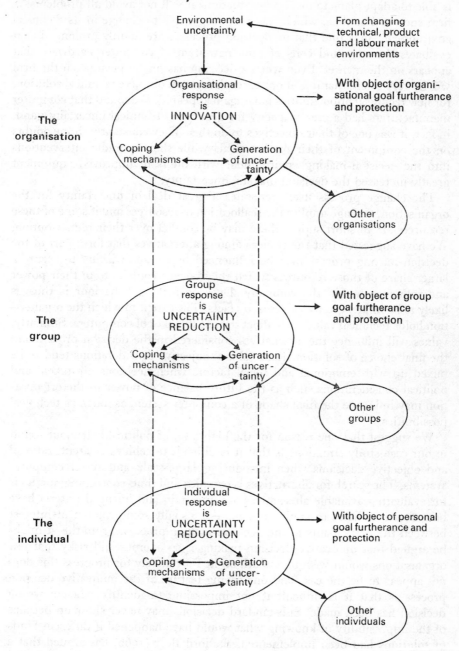

But even if the organisation is very responsive to environmental change and is able to adapt plans to meet new situations it will not avoid all problems. A firm such as Poultons, which was very responsive to change in its technical environment, may find that its decision processes are unduly prolonged as it evaluates the pros and cons of each new item of computer hardware that appears on the market. Even when a decision has finally been taken the firm will be very aware that it is about to implement an obsolete technical solution, for new technical opportunities have again appeared. We found that computer manufacturers had a great capacity for generating technical uncertainty and, in fact, it was one of their objectives to do this. They constantly threw doubts on the equipment of their market rivals while their sporadic interventions into the decision-making environment with offers of improved equipment greatly increased the decision makers' uncertainty.

The change process itself generates a great deal of uncertainty for the organisation. Change implies the reallocation of resources and if some of these resources are in short supply there may be conflict over their redistribution. We have suggested that the choices among alternatives that form part of the decision-making process may be influenced by groups wishing to secure a larger share of those resources which they believe contribute to their power and influence within the company. Decision-making behaviour is thus as likely to be influenced by values as by facts. The extent to which the organisation holds technical values will affect the use it makes of computers. Similarly, values will influence the alternatives considered in the decision process and the final choice of solution. Rational and objective considerations tend to be mixed up with emotional and social factors when decisions are taken, and political considerations such as the future balance of power in the organisation may influence the final shape of a computer system as much as technical possibilities.

We suggest that one reason for the high level of political behaviour found in our case study situations is that it is difficult to achieve clearcut, rational and objective decisions when introducing large-scale and novel computer systems. The search for alternatives is complex and time-consuming; methods for evaluating available alternatives are uncertain and financial criteria have been found difficult to apply, and there is usually some conflict of interest between the participants in the decision-making process. From this it might be argued that innovative decision-making is so onerous and risky that few organisations would wish to embark on it. Fortunately for progress this does not appear to be the case and one saving feature of the innovative decision process is that it is difficult if not impossible to identify when a wrong decision has been made. Substandard decisions may never show up because of the impossibility of knowing what would have happened if different kinds of solutions had been implemented. Stafford Beer (1966) has argued that a major objective for an organisation is 'improvement'. If a decision and its implementation move the organisation generally in the direction that it wants to go, then the decision has been a reasonable one and should not be criticised.

Uncertainty and the group

The organisation as a whole is affected by the need to recognise, understand and control uncertainty, but those parts of it associated with either the introduction or reception of the innovation are particularly affected. We have shown how both the specialist computer departments, and the user departments which receive the new system have to cope with major uncertainties during the change process. The way they cope and the manner in which they interact with each other greatly affect the ability of the organisation to assimilate the innovation easily, without too much strain and stress.

Groups and individuals have both organisational and personal interests when they cope with change. They seek to use the innovation to further organisational goals but they also recognise that the change situation can be used by themselves for personal advantage or by others for their disadvantage. Key individuals and groups in both computer and user departments may try to manipulate the change process to fit their own interests, or may believe that they have to protect themselves from attack by others who are engaged in this manipulation. This behaviour generates a great deal of uncertainty and leads to intense political activity in which strategies are developed to assist the achievement of personal and group goals or to provide protection from the threatening activities of others. The change situation is used by some participants in an expansionist manner to attain more power, influence, status, money, resources or whatever they perceive as important to them. It is also used in a maintenance manner by other, generally less powerful groups, to hang on to those elements of their situation which they cherish and to prevent the erosion of these by other groups.

Uncertainty may be increased by pressures in a group's internal environment caused by the need to assimilate something new, or by pressures in its external environment brought about by the need to work in close association with other groups while the innovation is being introduced. Uncertain situations tend to be both unpredictable and complex; that is, they contain many different variables all interacting with and influencing each other, although the nature and extent of this influence is not clearly understood by the group affected by the uncertainty. Some groups faced with uncertainty may react by pretending it is not there and doing nothing about it. A more normal response is to attempt to understand the reasons for the uncertainty and to seek to reduce it by securing control over all or some of the elements operating in the situation.

We would argue that a prerequisite to this control is *knowledge*. Unless situational variables are understood and can be manipulated by a group it cannot control them. It is because knowledge is related to control that the expert tends to be in an advantageous position when innovation is being introduced. If the innovation is a computer system the computer specialist will normally design the new system of work. He will be constrained in his design by the boundaries of his own knowledge and, unless challenged by other groups, by his own value position. Unless the user groups give a clear picture of their needs – and having such a picture implies considerable knowledge – they will get what the expert thinks they should have rather than what they might ideally like to have. User groups often try to combat this power of the expert by developing strategies to deal with it. If they see the new computer system as removing their own traditional areas of expertise these strategies may take the form of attempts to demonstrate that existing skills are still relevant, or of influencing the firm to be cautious and move slowly when introducing computer based technological change.

Controlling uncertainty located within a group's own environment is not easy, but controlling uncertainty which is a product of competitive relationships between groups with different needs, interests and objectives is even more difficult and there will frequently be a period of conflict before means for reconciling differences are worked out. If a group believes that it is threatened by another group or groups it is likely to draw together, to show a collective identity and to introduce group norms directed at emphasising group unity and solidarity. Each group is then likely to regard the other as an 'enemy'. Blake *et al.* (1964) suggest that negative actions feed upon each other; if one group acts with hostility towards another then this will provoke a counter action. If two competing groups have an interdependent relationship – that is, they are unable to work in isolation but depend on each others' services for the successful completion of a task – then hostility is likely to be increased for they are forced into a constant and irritating association.

A major problem for any expert group seeking power and control as a means for achieving its goals is to establish its legitimacy. It must demonstrate that the knowledge it has is crucial to the successful performance of other groups and give them confidence in the level and relevance of the knowledge which the specialists possess. Only if a group is perceived as the legitimate possessor of a certain knowledge area will it be able to retain its power. But, unless there is a very considerable disbalance of power, groups interacting together within a change situation cannot control each other but have to use *influence* instead. This implies developing strategies which guide the change situation in a direction which achieves the group's interests without evoking opposition from other groups. Such strategies involve negotiation and the formation of coalitions in an attempt to gain support for a particular interest position.

Groups associated with the introduction of innovation, if they are to work together at all, must have ways of reducing conflict to ensure that it does not become so dysfunctional that it reduces work performance. In most situations norms and sanctions which act as regulatory mechanisms and keep conflict at a manageable level develop over time. But this is a slow process and conflict is often a feature of the early stages of innovation when technical specialists and users are working together for the first time. However, if the change is to be successfully introduced the groups must at some point decide on a set of mutually desired objectives and work cooperatively towards these, and in this book we have described some of the problems that have to be overcome before this cooperation can be achieved.

The three groups most commonly in association when large-scale computer systems are introduced – top management, computer specialists and the management and staff of the user department will all experience some degree of uncertainty and stress. Top management's uncertainty is likely to be related to its willingness and ability to control computer innovation. Some Boards of Directors incorporate plans for new computer systems into corporate strategy and monitor very closely the successes and failures of this technology. Others leave a great many of the decisions concerning the use of computers to their technical specialists and merely provide sanction and the necessary finance. The Board is always in a risk relationship to new computer systems but it does not usually experience the degree of stress and uncertainty felt by groups which have to implement or operate the new system.

Computer specialists have the stress of meeting time targets and deadlines and the uncertainty of not knowing if their decisions are the right ones, but they also have the security of being in control of the design process and of having technical competence – the security of 'knowledge'. The group with the greatest amount of uncertainty and stress is likely to be the user department. It will have to play a major part in implementing the new system and will be held responsible for its successful operation.

Uncertainty as a product of risk

It is useful to summarise the risks inherent in the introduction of computer systems, as risk is an important generator of uncertainty. Some risk factors will affect all groups in the innovation situation, others will affect one group but not others. Whether a group experiences a particular risk or not is likely to be related to how it defines its role and responsibilities in the change situation.

FINANCIAL RISK. The risk of heavy financial losses through a capital investment which does not pay for itself will be carried by the organisation as a whole although the Board will assume responsibility for avoiding this kind of catastrophe. The risk is likely to be greater if the firm's environment is unstable and the product market precarious. If the firm is in a healthy economic situation with growing markets then the costs of a technical failure will be more easily absorbed. When innovation is being introduced in a difficult economic period, pressure on computer specialists and user departments is likely to be considerable. They will be required to avoid failure and bring the new system into operation as soon as possible so as to achieve a return on capital. Both these features of the change situation will produce stress, anxiety and uncertainty.

OPERATIONAL RISKS. If the computer system is a novel one, designed to tackle an entirely new problem or an old problem in a new way, the firm is likely to be in a research and development situation and to have some doubts about how successful the new system will prove to be. This kind of situation will generally cause top management to observe and control the activities of its computer specialists carefully, and will cause the computer specialists, in turn, to monitor the user department closely. If the new system is not entirely successful then scapegoating may occur, with the computer specialist attempting to convince top management that problems are due to user incompetence.

HUMAN RELATIONS RISKS. All change has human consequences and computers are no exception. A new system may fundamentally alter the skill structure of the user department and the number of staff required there. Changes of this kind have the potential to produce serious industrial relations problems.

POLITICAL RISKS. A major thesis of this book and the earlier work by Pettigrew has been that major technical change implies the reallocation of resources; a process in which some groups gain and others lose if these resources are in short supply. Resources can be material items such as equipment, but they can also be money, skills, services and other tangible and intangible things which give the holder status, influence and power. Technical change in our case study firms was characterised by groups attempting to gain additional status, power and influence through the acquisition of more resources than they had before the change and by other groups attempting to hold on to what they already possessed. We have described how the process of doing this produced a high level of political activity both within computer departments and between computer specialists and user management.

The response to uncertainty

The acceptability of technical change seems often to be related to the extent to which user departments are able to participate in decision processes concerning the proposed installation. This participation is a consequence of the allocation of responsibilities between computer specialists and users and this allocation, in turn, is a product of the way each group defines its role. The user manager who is knowledgeable about computer systems, and can identify his own problems and appreciate which of these lend themselves best to a computer based solution, is likely to demand and get a high level of participation. We found that user managers in both AEK and Falcon Ltd participated in the design of a new system and took considerable responsibility for its implementation and operation. In contrast many user managers are forced to assume a more passive role because they lack knowledge and competence in what to them is a new technical area. In Poultons and Grant and Co. user managers could play little part in the planning and design of their new computer systems. This increased their stress and uncertainty because they were not in control of their own situations and yet were likely to be held responsible for the successful operation of the new systems.

We have seen that computer specialists differ considerably in their approach to user departments. Some see themselves in a 'service' role and wait until line managers approach them with problems. Others are more actively engaged in seeking out problems, see themselves in a 'ginger' role and believe that one of their most important functions is to develop more innovative attitudes and interests in other parts of the firm. This choice of role to some extent reflects the interests and motivation of powerful individuals and groups within the computer department, but it is also related to the nature of the situations in which the computer specialists find themselves. For example, in Grant and Co. the computer department was forced to change from a 'service' to a more active role when it became clear that the user department could not handle a technical change that was entirely new to them. Before this change of role took place user management in this firm were under very considerable stress for they were placed in a controlling position without the knowledge and competence to exercise this control.

We believe that a mutually agreed allocation of responsibilities between technical specialists and user groups takes time and that in most situations a process of bargaining for different areas of responsibility will take place. In the initial stages of any innovation this bargaining process will be clouded by a lack of understanding of the nature of the tasks available for allocation. There is likely to be a great deal of taking up, dropping and transferring of task responsibilities as each group attempts to define its activities in terms of its competence, needs and objectives. Kahn *et al.* (1964) point out that when this allocation process is difficult to resolve conflict is liable to break out between the groups. Some specialists and users will react to this conflict with aggression and make attempts to stake their claims forcibly, others will try to protect themselves from a stressful situation by withdrawing and leaving others to take the decisions.

Lack of knowledge as a factor in uncertainty

Both computer specialists and user departments can suffer uncertainties because of an absence of timely and relevant knowledge, but in most situations the technologist is in a better position to fill this knowledge gap. A search for information will be an integral part of the planning and decision stages of a major computer system and he will have the assistance of competing computer manufacturers in making this search. We have referred to the fact that in the computer field a mating process takes place, with problems looking for solutions and solutions seeking out problems. The user department is likely to have greater difficulty in filling its knowledge gaps and to lack the resources to do this. In addition there may be little enthusiasm on the part of the computer specialist to assist the user department with its knowledge problem, as this could lead to a transfer of control from one group to the other. This lack of knowledge is likely to place a number of different stresses on the user manager and his department. Because they are not involved in the design of the new computer system they are likely to be anxious that they will not understand the new methods of work when they are introduced.

The user manager will have particular knowledge problems at this time. He is required to be a 'change agent' and to transform his department from one technical state to another – a process which usually involves a major reorganisation of work procedures. The process of 'changing' is itself both skilled and difficult. The user manager is likely to experience pressure from top management to get the new system operational and efficient as soon as possible, yet the successful assimilation of technical change involves the development of new attitudes and skills in the group using the new system, and it will be the user manager's responsibility to ensure that his staff acquire these. To achieve success he will need to be a good diagnostician of the human problems associated with the change process and an excellent communicator to his staff. Many managers will have difficulty in both these areas and will be under stress because they can meet neither the expectations of top management nor those of their own staff.

The need to reconcile different interests

We have already referred to the fact that the different groups involved in innovation are unlikely to have a complete identity of interest. In some situations they will have major conflicts of interest and these will have to be recognised and reconciled during the change process. The technical specialists will be keen to optimise the use of the technology which they know and understand, and this can lead them to design systems which have a high technical competence but are poor at catering for human needs, such as a desire for job satisfaction. The user group is unlikely to be able to challenge the technical knowledge of the specialists and this can force it into a dependency relationship. In order to combat this power of the 'expert' the user department will have to generate support for its own alternative demands. Invariably this support has to be sought from top management and can only be achieved by throwing doubt on the validity of the 'expert' proposals. This must inevitably lead to a conflict situation in which the experts will come off best if they are seen by top management as assisting the firm to cope with uncertainties and pressures originating in its environment. The power of the expert is greatest when he is innovating in an area critical to the survival of the organisation.

The specialist, if he is to operate effectively in an organisation, must reduce conflict by finding out how other groups see his responsibilities and establishing the extent to which these coincide with his own role definition. If he finds that there are major differences between the way he is defining his role and the way it is being defined by others then he has the problem of how to reconcile these differing expectations. One source of difficulty in achieving agreement on role definition is the divergence in objectives of specialist and line departments. The survival of computer specialists as a viable group depends on their ability to press forward with innovation. In contrast, user managers may give their production responsibilities top priority, because their performance is evaluated in terms of their ability to meet production targets and delivery dates. User departments require convincing that the activities of computer specialists are in line with, rather than against, their own interests.

When specialists belong to an external group sent into a local firm by the head office of a parent company they are likely to encounter considerable difficulties. They will have less identification with local problems and needs than the local men, yet their success and acceptance may depend on their ability to involve themselves completely and sympathetically with the needs of the local establishment.

The eventual and necessary reconciliation of interests will be easier if the technical expert interprets his role as a facilitator of change rather than simply as an agent of high level technology. This implies an ability to recognise the human relations factors in the change situation and a willingness to abandon some technical goals in order to achieve human goals. We have suggested that a variety of factors influence how the computer specialist defines his role. If he is a long service employee, recruited into the computer department from another less technical part of the firm, he is likely to be well indoctrinated into company values and to be conscious of the non-technical needs of the user department. If, however, he is a technical man, perhaps in his first technical job after university, he is likely to define his job in purely technical terms and to be unable to take account of human factors. He may have difficulty in understanding the user manager's concern for human relations and department morale, or even the very real business constraints under which he has to operate.

A narrowly defined role and set of responsibilities reduce the computer specialist's uncertainty by limiting the number of variables of which he has to take account. But it is likely to increase the uncertainty of the user department which may feel that its interests are being overlooked.

Conflicts of interests and political behaviour

Attempts to use the innovation process to achieve group objectives or safeguard group interests cause political behaviour. There are battles over scarce resources such as money and power, with both specialists and users seeking to use the change situation to promote their own interests and secure more influence and control within the organisation. We have already indicated that innovation alters existing patterns of resource sharing. New resources may be created by the innovation and come within the jurisdiction of a department or group which has not previously had this kind of resource. The group may then perceive these resources as an opportunity for gaining status and power relative to others in the organisation. At the same time those who see their interests as threatened by the proposed change may resist such allocation of new resources. In this way political behaviour is generated.

Pettigrew (1973b) argues that a great deal of the uncertainty and stress associated with a major innovation is related to the claims each side makes for its knowledge and skills to be regarded as important resources. The computer specialists are likely to argue that there must be increased recognition of the value of technical information as a business resource. The user manager may argue that his department's knowledge of customer needs and of the product market is of greater value than sets of computer produced statistics. He may also fear that if the claims of the computer specialists are accepted by top management this will lead to his department's skill and competences being undervalued and to his losing status and influence within the organisation.

Coping with uncertainty

Both computer specialists and user departments will develop mechanisms to handle the uncertainties with which they are presented during the change process. We have seen that the two groups have problems of knowledge and interpersonal relationships. The computer specialists' technical knowledge problems will never be completely solved because the technical environment in which they are operating is too dynamic. They will however become increasingly experienced and able in searching out new information, in evaluating this and in choosing between the various alternatives that have been identified. A more difficult and critical area is the establishment of cooperative relationships with user departments and here a critical factor will be the specialist group's definition of its role and responsibilities. New occupational roles always present their occupants with problems. They have no history and therefore no agreed set of rights and duties has become associated with them.

During a time of major change both computer specialists and user managers are prone to suffer from role strain and role uncertainty because neither they, nor the groups with which they interact, understand precisely what their particular skills are and how to use them. Top management may not provide the computer specialists with a clear definition of the tasks they are to perform because it has not thought through the nature of the specialist role in any meaningful way. Because of this the computer specialists may try to usurp aspects of the change process which the user department regards as coming within its responsibilities. Again, the user department may not have had any previous association with computer specialists and be very unclear concerning where their responsibilities begin and end. Situations in which the computer specialists are defining their role in one way and the user department is defining it in another will be stressful for both groups and there is likely to be a great deal of negotiation before the allocation of tasks is successfully completed.

Role definition, like all other elements of our innovation process, is affected by political factors. Agreement on the allocation of tasks between different groups may be partly arrived at on the basis of knowledge and competence but it will also be influenced by each group attempting to reduce the uncertainty of the change process through shaping it to meet its own interests. The specialist group may favour a broad role definition that increases its ability to control the change situation and reduces the power of the user department to interfere. The ability to do this requires either a weak, compliant or passive user department or approval from top management of the computer specialists' right to define their role in this way. Where this imbalance of power does not exist an important factor in role definition will be what Turner (1962) has called 'feedback testing transactions'. Each group will challenge the activities of the other and seek to take some of these over. This compels the first group to rethink and modify its activities in order to maintain a working relationship with the second group.

If a user manager is to retain control of his own situation and to influence the outcome of the change he needs knowledge as well as power. He must be able to insist on participating in design and planning decisions and competent to represent his department's interests when he does participate. Effective participation usually requires some experience in using computer systems and is not easy for the user manager who is new to this form of technical change.

Successful role definition may assist the avoidance or reduction of conflict associated with the allocation of task responsibilities, but it contributes little to conflicts of interest over needs and objectives. To solve this kind of problem there must be a willingness to recognise the existence of conflict and to bring it out into the open. A reconciliation of these different interests has to take place, with each group being willing to abandon some of its personal goals in the hope of achieving a result acceptable to everyone. An unwillingness to resolve conflict may arise because either the computer specialist group or the user department believes that it can actively guide the change process in the direction of its interests and does not want a compromise solution. When this situation arises conflict is inevitable and may be irresolvable. The situation can only be eased through the intervention of a higher level group such as top management which steps in and prevents this suboptimisation of organisational goals. We would support Cyert and March (1963) in believing that conflict resolution can only be partially successful and that 'most organisations, most of the time, exist and thrive with considerable latent conflict of goals'. During periods of major change it is probable that latent conflict will become overt conflict.

Uncertainty and the individual

Individuals experience the same difficulties as groups. They have to reconcile organisational and personal goals; they have to understand the dynamics of the change situation if they are to cope with its complexity and gain a measure of control over it; they have to accommodate the interests of other individuals

whose objectives are not identical with theirs. In chapter 5 we gave some examples of how individuals handled this complex situation and how they used
the internal political environment as a means for achieving their goals and protecting their interests. An individual's personal goals will be related to his
psychological and social needs. He is likely to be seeking influence, status and
psychological security, together with opportunities for self development and
achievement. His family responsibilities will mean that he also requires an
adequate financial reward, opportunities for promotion and job security.
His psychological needs will be met or frustrated through the nature and task
content of his occupational role and we have already discussed how occupational roles have a tendency to fluidity during periods of change when the
allocation of tasks can affect the ability to pursue personal and group interests.
If he is a computer specialist the speed of development in the technological
environment will alter his role and require him to acquire new competences.
If he is a user manager the problem of coping with innovation may require
him to develop new skills and knowledge areas.

The individual in a boundary role has to interact with groups other than
his own and this requires credibility and legitimacy. His success, and the
realisation of his personal goals, may depend upon the respect in which he is
held, upon his personal acceptability and upon the extent to which others feel
indebted to him for past services. If he requires the support of others to
achieve his objectives then the timing and manner in which he seeks this
support may have a crucial impact upon the success which he achieves
(Pettigrew, 1973b).

Individuals use political means to achieve their objectives. They seek to
influence those people who are associated with the kind of resource which
they seek and value, and they attach themselves to groups which they perceive as having power in those areas which are critical to the realisation of
their goals. Burns (1965) tells us that positions of power offer their occupants
both material and psychological rewards and this itself stimulates conflict
between those who have power and those who want it. This desire for power
can lead to the development of strategies directed at coercion or manipulation
of others.

Barnard (1938) suggests that a kind of dual personality is required of
individuals contributing to organisational action – a private personality
and an organisational personality. A key problem for individuals in both
specialist and user groups when coping with the problems of major change is
developing strategies which enable them to achieve both personal and organisational goals. The reconciliation of these two sets of interests can depend
on their ability to influence the attitudes and behaviour of others through the
possession of some scarce resource which they possess. If the individual is a
technical specialist this is likely to be knowledge; if he is a user manager it will
be cooperation; if he is a top manager it will be power. For this scarce resource
to be a useful lever in assisting the individual's interests, it must be seen by
others as critical to the successful introduction of the innovation.

Innovation and uncertainty

The organisation, the group and the individual are all required to cope with different kinds of uncertainty when introducing a major technical innovation such as a large-scale computer system. This uncertainty stems primarily from three features of the change situation – an absence of knowledge, an exposure to risk and the need to compete for scarce resources which are being reallocated by the change process. The organisation tackles its lack of knowledge by creating groups of specialists such as computer technologists and giving them responsibility for the information search, evaluation and choice process. The values of the organisation influence the manner in which these groups approach this task and we have shown how different approaches to planning and decision-making were used in our firms. We have also shown how this knowledge gap is never satisfactorily solved, for as one set of information is mastered, technical developments and the dynamics of the change process itself produce new information needs.

The computer specialists see their responsibility as seeking out technical information although it can be argued that their unwillingness or inability to take account of human variables leads to systems design which concentrates on technical and ignores human needs. The user department is left to protect its own human relations situation, but the power to do this may be taken away from it by the way decisions are allocated between the two groups. The user department also has the knowledge requirement of introducing the new system, understanding it and learning to operate it. An inability to do this at the required speed may place it under severe stress.

Uncertainty which stems from the risks of major change is also experienced at each organisational level. The organisation as a whole will lose financially if a major technical investment proves unsuccessful. It will also be left with those difficulties and pressures in its environment which the innovation was introduced to relieve. The computer specialist group runs the risk of losing credibility and influence if its new system does not meet expectations. The justification of its existence is successful innovation and failure may put its survival in jeopardy and will certainly affect the degree of power and autonomy which the organisation allows it to have. The user department is subject to many risks. It will be seen as incompetent and resistant to change if it cannot assimilate the new system easily and efficiently, yet the stresses of doing this may be considerable. User staff may feel threatened by the possibility of redundancy and fearful of the need to learn new skills. User managers may be particularly vulnerable at this time for the computer specialist group may exclude them from the decision processes yet hold them responsible for the system's successful operation.

Change leads to the reallocation of resources, and when they are desirable and in short supply there will be intense competition for a major share in these. Resources likely to be most sought after are power, influence, knowledge, money and status, all of which enable the organisation and the groups and individuals within it to exert control over their environments and in doing this to move towards the achievement of their goals. The organisation hopes that its new computer system will give it greater control over its product market environment by enabling it to serve customers faster and more efficiently.

Fig. 9.2 Innovation decision-making when the balance of power is equal

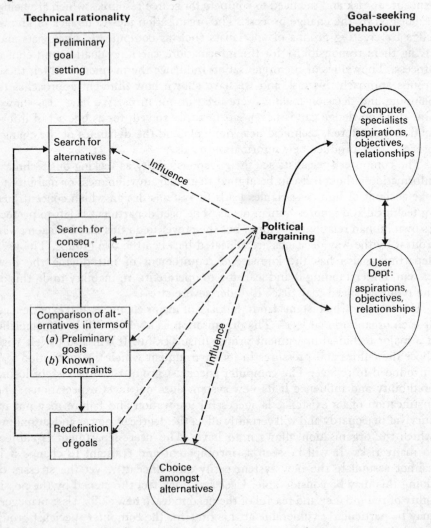

Computer specialists and user departments will seek to use the fluidity of the change situation to obtain more of those resources which are important to them and to prevent the removal of these by other groups. Individuals will attempt to achieve personal goals and protect personal interests while assisting the organisation to achieve its goals. Because the interests of different groups and individuals do not coincide, competition for desired resources will lead to intense political behaviour, with each faction attempting to guide the change process and its consequences in the direction which suits its interests. If left unchecked this competition will lead to major conflict which can threaten the stability of the organisation. It has to be handled through good negotiating mechanisms and in all of our firms these were formalised through the creation of committees. These committees had the function of bringing conflicts of interest into the open where they could be discussed and resolved. They also assisted agreement on how resources should be shared through the manner in which they allotted responsibilities for different aspects of the change process. We found that role definition and the distribution of task responsibilities were important means for giving a group control over desired resources.

Figure 9.2 illustrates how the mechanisms used by our four firms to deal with uncertainty of innovation were a mixture of technical rationality, goal-seeking behaviour and political bargaining.

The variety reduction process of searching for information, evaluating this and choosing among alternatives had a high rational content, but it was subject to political pressures arising from the goal-seeking activities of the different groups. In those firms which had structured planning and decision-taking mechanisms political bargaining to reconcile goals and interests took place mainly in committees, and was then fed into the technical decision processes. In the firms which gave a great deal of power and responsibility to their computer specialists, political bargaining could occur within the computer department and the different goals of the technologists had to be reconciled before final decisions were taken.

Planning for change as an adaptive process

A consistent thread running throughout this book has been the idea that planning will be an imperfect process. In earlier chapters we have raised detailed and specific doubts about notions of rationality associated with planning and questioned those who assume planning is more of an act than a process. In terms of Ackoff's (1970) recent typology of planning we have argued that the rationality and process associated with *optimising planning* is likely to be inefficient in many change situations and that *satisficing planning*, the philosophy of 'doing well enough' rather than as well as possible, seems more realistic and attainable. Our goal however, is closer to Ackoff's third philosophy of planning; adaptive or innovative planning.

Innovative planning recognises that the process of producing plans is rather more important than the plans themselves. The outcome is subsumed by the process. It is participation in the process, not the consumption of the product, which is critical, not only because the nature of the process of production will crucially affect the degree of commitment to the plan but also because the process is an important mechanism for learning to learn, and without this no system under change could hope to develop the adaptive capacity to cope with future organisational uncertainties.

Some of the data from our case-study firms is suggestive of maladaptive planning. Emery *et al.* (1974) define maladaptivity as a reactive or defensive mode of meeting uncertainty, where the organisational turbulence is ignored rather than actively confronted. Thus choice processes to go ahead with an innovation may take the form of win–lose encounters when natural lines of historical association and dissociation are mobilised and where cooperation is limited to those people, or that knowledge, the participants know and are comfortable with. Or there may be acts of denial by the innovators themselves, the belief that users have little to contribute in the change process and the associated indifference to the client's needs, and when these ways of coping with uncertainty bring their unanticipated fruits, there is superficiality; the innovator and user lower their emotional investment in the change being pursued.

For Emery *et al.* and for the authors of this book, this way of approaching organisational innovation is passive and maladaptive. Passive because it is only the presenting symptom and surface manifestation of uncertainty which is approached and maladaptive because the causes of uncertainty lie undisturbed and yet more uncertainty has been released by the original reactive strategy. The interactive process of *uncertainty providing the enabling conditions for internal politics and the expression of political behaviour creating further uncertainty* is basic to our whole concept of the dynamics of innovation within the firm. Our theory of changing recognises internal politics, the acquisition and maintenance of power and status by individuals and groups, to be a natural and consequential part of the change process. One active and seemingly adaptive way of planning for change therefore is to anticipate forms and sources of political activity and to take a proactive interest in those political processes.

Whereas passive maladaptive planning as a way of handling uncertainty is dominated by the psychological and system state of approach – avoidance or evasion, active innovative planning is a process concerned with the understanding and modification of organisational uncertainties. These uncertainties are by no means unidimensional. One of Emery *et al.*'s criticisms of planners is that they assume that we simply need to know more and more facts when what is really needed is knowledge of values. Our research indicates that some of the key uncertainties surrounding organisational innovation are about technical knowledge, role definitions, work styles, political interests and values. Each of these sources of uncertainty may find expression in one or other of a number of levels of analysis. Some will appear as individual concerns and opportunities, some as group phenomena, others as organisational factors, and still others at all levels of analysis, as the interaction between the different dimensions of uncertainty repercusses throughout the system.

This research has also established that the diagnosis of uncertainty, the first stage of innovative planning, requires a temporal perspective. While planning in its very nature is an anticipatory and futuristic activity, if the plans are to be implemented successfully they should be soundly based in the past and present as well as the future. People's reactions to innovation are likely to be a complex mixture of, 'Can I connect this proposed change to what has gone before, to the constraints and opportunities before me now, and to the likely impact of events on me in the future?' Innovative planning is temporal not only in the sense that it is not a once and for all activity but also in its analytical mode.

But diagnostic skills are only one part of the planning process. The process of developing and producing the plans is critical. This covers customary modes of conflict resolution or consensus seeking, of mutually exploring definitions of role between user and innovator and of bargaining over the re-distribution of political resources? Here our work has reinforced the findings of Bass (1970) and Taylor and Irving (1971) about the consequences for the commitment to, and implementation of plans of the separation of the planners and the doers. But as Emery *et al.* have suggested, the real intangibles in the planning process are to do with disagreements and discrepancies about values. Values signify what information is more or less useful, what persons are included in the community of planners, what style of working is accept-able within the culture of the organisation undergoing change and perhaps most significantly what ends the innovation is supposed to serve. Attention to these factors is probably at the core of any adaptive response to organisational uncertainty which results from innovation and change.

References

Abbreviations:
Admin.Sci.Q.   *Administrative Science Quarterly*
Am.J.Sociol.   *American Journal of Sociology*
Am.Sociol.Rev.   *American Sociological Review*
Hum.Rel.   *Human Relations*
J.Bus.   *Journal of Business*

ACKOFF, R. L. (1970) *A Concept of Corporate Planning*, Wiley.

ALLEN, T. J. (1965) 'Problem solving strategies in parallel research and development projects', Working Paper no. 126-65, Cambridge, Mass., Sloan School, MIT.

ANTHONY, R. N. (1965) *Planning and Control Systems: a framework for analysis*, Harvard University Press.

APPLEY, M. H. and TRUMBULL, R. (1967) *Psychological Stress*, Appleton.

ARGYRIS, C. (1970) *Intervention Theory and Method: a behavioral science view*, Addison-Wesley.

ARGYRIS, C. (1971) 'Management information systems: the challenge to rationality and emotionality', *Management Science*, **17**, no. 6, B-275-292

ARGYRIS, C. (1972) *The Applicability of Organizational Sociology*, Cambridge University Press.

ARROW, K. J. (1951) *Social Choice and Individual Values*, Wiley.

BAILEY, F. G. (1969) *Stratagems and Spoils: a social anthropology of politics*, Oxford, Blackwell.

BAKER, J. K. and SCHAFFER, R. H. (1969) 'Making staff consulting more effective', *Harvard Business Review*, Jan.–Feb., pp. 63-71.

BANNESTER, D. M. (1969) 'Socio-dynamics: an integrating theorem of power, authority, interinfluence and love', *Am. Sociol.Rev.*, **24**, no. 3, 374-93.

BARNARD, C. I. (1938) *The Function of the Executive*, Harvard University Press.

BASS, B. B. (1970) 'When planning for others', *Journal of Applied Behavioural Science*, **6**, no. 2, 151-71.

BAUER, R. A. (1966) *Social Indicators*, MIT Press.

BAUMAN, Z. (1967) 'The limitations of "perfect planning" ', in Gross (1967).

BECKER, H. S. and CARPER, J. W. (1956) 'The development of identification with an occupation', *Am.J.Sociol.*, **61**, no. 4, 289-98.

BEER, S. (1966) *Decision and Control*, Wiley.

BEER, S. (1969) 'The aborting corporate plan – a cybernetic account of the interface between planning and action', in E. Jantsch, ed., *Perspectives of Planning*, Paris, OECD.

BEER, S. (1972) *The Brain of the Firm*, Allen Lane, The Penguin Press.

BENNIS, W. (1969) *Organization Development: its nature, origins and prospects*, Addison-Wesley.

BLAKE, R. R., SHEPARD, H. A. and MOUTON, J. S. (1964) *Managing Intergroup Conflict in Industry*, Houston, Gulf Publishing.

BLANKENSHIP, L. V. and MILES, R. E. (1968) 'Organisational structure and managerial decision behaviour', *Administrative Science Quarterly*, **13**, no. 1, 106–20.

BLAU, P. M. (1964) *Exchange and Power in Social Life*, Wiley.

BRAYBROOKE, D. and LINDBLOM, C. E. (1963) *A Strategy for Decision: policy evaluation as a social process*, Free Press of Glencoe.

BROSS, I. D. J. (1953) *Design for Decision*, New York, Macmillan.

BROWN, W. (1960) *Exploration in Management*, Heinemann.

BRUNER, J. S. and POSTMAN, L. (1949) 'On the perception of incongruity: a paradigm', *Journal of Personality*, **18**, 206–23.

BUCHER, R. (1970) 'Social process and power in a medical school', in N. M. Zald, ed., *Power in Organizations*, University of Vanderbilt Press.

BUCKLEY, W. (1967) *Sociology and Modern Systems Theory*, Prentice-Hall.

BURNS, T. (1965) 'On the plurality of social systems', in J. Lawrence, ed., *Operational Research and Social Sciences*, Tavistock Publications.

BURNS, T. and STALKER, G. M. (1961) *The Management of Innovation*, Tavistock Publications.

CADWALLADER, M. L. (1969) 'The cybernetic analysis of change', *Am.J.Sociol.*, **65**, 154–7.

CHURCHMAN, C. W. (1968) 'The case against planning – the beloved community', *Management Decision*, Summer, pp. 74–7.

COOKE, J. E. and KUCHTA, T. (1970) 'Feasibility studies for the selection of computer systems and applications', *Cost and Management* (Canada), Sept./Oct., pp. 11–19.

COSER, L.A. (1964) 'The termination of conflict', in W. J. Gore and J. W. Dyson, eds, *The Making of Decisions*, Free Press of Glencoe.

COTTRELL, F. (1962) 'Social groupings of railroad employees', in S. Nosow and W. H. Form, eds, *Man, Work and Society*, New York, Basic Books.

CROZIER, M. (1964) *The Bureaucratic Phenomenon*, Tavistock Publications.

CYERT, R. and MARCH, J. (1963) *A Behavioral Theory of the Firm*, Prentice-Hall.

CYERT, R. M., DILL, W. R. and MARCH, J. C. (1967) 'The role of expectations in business decision making', in M. Alexis and C. Z. Wilson, eds, *Organizational Decision Making*, Prentice-Hall, pp. 34–47.

CYERT, R. M., SIMON, H. A. and TROW, D. B. (1956) 'Observation of a business decision', *J.Bus.*, **29**, 237–48.

DAHL, R. A. (1957) 'The concept of power', *Behavioural Science*, **2**, 201–18.

DALTON, M. (1959) *Men Who Manage*, Wiley.

DAVIES, D. and MCCARTHY, C. (1967) *Introduction to Technological Economics*, Wiley.

DEARBORN, D. C. and SIMON, H. A. (1958) 'Selective perceptions: a note on the departmental identification of executives', *Sociometry*, **21**, 140–4.

DEUTSCH, M. (1969) 'Conflicts: productive and destructive', *Journal of Social Issues*, **25**, 7–41.

DEVONS, E. (1950) *Planning in Practice*, Cambridge University Press.

DILL, W. R. (1958) 'Environment as an influence on managerial autonomy', *Admin. Sci.Q.*, **2**, no. 4, 409–43.

DOWNS, A. (1967) *Inside Bureaucracy*, Little, Brown.

DUFTY, N. F. and TAYLOR, P. M. (1962) 'The implementation of a decision', *Admin. Sci.Q.*, **7**, June, pp. 110–19.

DUNCAN, R. B. (1970) 'A cybernetic-operant model of organizational learning: an exploration of how organizations learn to adapt to the uncertainty in their environment', unpublished paper, Yale University.

DUNCAN, R. B. (1972) 'Characteristics of organizational environments and perceived environmental uncertainty', *Admin.Sci.Q.*, **17**, no. 3, 313–27.

DUNCAN, R. B. (1973) 'Multiple decision-making structures in adapting to environmental uncertainty: the impact of organizational effectiveness', *Hum.Rel.*, **26**, no. 3, 273–91.

EASTON, D. A. (1965) *A Systems Analysis of Political Life*, Wiley.

ELLUL, J. (1965) *The Technological Society*, Cape.

EMERSON, R. M. (1962) 'Power-dependence relations', *Am.Sociol.Rev.*, **27**, 33–41.

EMERY, F. E., ed. (1969) *System Thinking*, Penguin Books, Modern Management series.

EMERY, F. E. and TRIST, E. L. (1965) 'The causal texture of organisational environments', *Hum.Rel.*, **18**, 21–32.

EMERY, F., EMERY, M., CALDWELL, G. and CROMBIE, A. (1974) 'Futures we're in', unpublished paper, Australian National University, Centre for Continuing Education.

EMERY, J. C. (1969) *Organisational Planning and Control Systems: theory and technology*, Collier-Macmillan.

ETZIONI, A. (1968) *The Active Society*, Collier-Macmillan.

FRENCH, J. R. P. and RAVEN, B. (1966) 'The bases of social power', in D. Cartwright, ed., *Studies in Power*, 2nd edn, Ann Arbor, Michigan, Institute for Social Research.

FRIEDMANN, J. (1967) 'The institutional context', in Gross (1967).

FRISCHMUTH, D. S. and ALLEN, T. J. (1968) 'A model for the description and evaluation of technical problem solving', Working Paper, Cambridge, Mass., Sloan School, MIT.

GALBRAITH, J. (1953) *Designing Complex Organizations*, Addison-Wesley.

GARNER, W. (1962) *Uncertainty and Structure as Psychological Concepts*, Wiley.

GLUCKMAN, M. (1956) *Custom and Conflict in Africa*, Oxford, Blackwell.

GOETZ, B. E. (1949) *Management Planning and Control*, McGraw-Hill.

GOFFMAN, E. (1969) *The Presentation of Self in Everyday Life*, Allen Lane, The Penguin Press.

GOODE, W. J. (1960) 'A theory of role strain', *Am.Sociol.Rev.*, **25**, no. 4, 483–496.

GOULDNER, A. W. (1957) 'Cosmopolitans and locals: towards an analysis of latent social roles', *Admin.Sci.Q.*, **2**, 281–306.

GROSS, N., MASON, W. S. and MCEACHERN, A. W. (1958) *Exploration in Role Analysis*, Wiley.

GROSS, B. M. (1967) *Action Under Planning*, McGraw-Hill.

HALL, E. T. (1968) *The Silent Language*, New York, Fawcett.

HALL, R. H. (1962) 'Intra-organizational structure variation', *Admin.Sci.Q.*, **7**, 295–308.

HARVEY, E. (1968) 'Technology and structure of organizations', *Am.Sociol.Rev.*, **33**, 247–59.

HAWGOOD, J. and MUMFORD, ENID (1971) 'Decision strategy', in W. E. M. Morris, ed., *Economic Evaluation of Computer Based Systems*, Book 1, Manchester, National Computing Centre.

HEDBERG, B. (1974) 'Systemutformning som en forandrings-process. En socioteck-nisk referensram', in I. Widlund and B. Fristedt, eds, *Processkonsultation*, Stockholm, Allmanna Forlaget.

HICKSON, D. J., HININGS, C. R., LEE, C. A., SCHNECK, R. E. and JENNINGS, J. M. (1971) 'A strategic contingencies theory of intra-organisational power', *Admin.Sci.Q.*, **16**, no. 2, 216–29.

HOMANS, G. C. (1965) *The Human Group*, Routledge.

HOMANS, G. C. (1967) *The Nature of Social Science*, Harbinger.

HUNT, R. (1968) Review of E. J. Miller and A. K. Rice, *Systems of Organization*, in *Admin.Sci.Q.*, **13**, 360–2.

KAHN, R. L. and QUINN, R. P. (1968) *Role Stress: a framework for analysis*, University of Michigan, Survey Research Centre.

KAHN, R. L., WOLFE, D. M., QUINN, R. P., SNOEK, J. D. and ROSENTHAL, R. A. (1964) *Organizational Stress*, Wiley.

KAPFERER, B. (1969) 'Urban Africans at Work', unpublished PhD thesis, University of Manchester, Department of Social Anthropology.

KATZ, D. and KAHN, R. L. (1966) *The Social Psychology of Organizations*, Wiley.

KNIGHT, F. (1921) *Risk, Uncertainty and Profit*, Harper & Row.

KNIGHT, K. E. (1967) 'A descriptive model of the intra-firm innovation process', *J.Bus.*, **40**, no. 4, 478–96.

KOOPMAN, B. O. (1956) 'The theory of search 1', *Operational Research*, **4**, 324–46.

LAWRENCE, P. R. and LORSCH, J. W. (1967) *Organization and Environment*, Harvard Business School, Division of Research: Harvard University Press.

LAZARUS, R. S. (1966) *Psychological Stress and the Coping Press*, McGraw-Hill.

LEAVITT, H. J. (1958) *Managerial Psychology*, University of Chicago Press.

LI, D. H. (1972) *Design and Management of Information Systems*, Chicago, Science Research Association Inc.

LINTON, R. (1965) *The Study of Man*, Peter Owen.

LYNTON, R. (1969) 'Linking an innovative sub-system into the system', *Admin.Sci. Q.*, **14**, no. 3, 398–416.

LUCE, R. D. and RAIFFA, H. (1957) *Games and Decisions*, Wiley.

MCCLEERY, R. (1960) 'Communication patterns as bases of systems of authority and power', in *Theoretical Studies in Social Organization of the Prison*, New York, SSRC pamphlet.

MCGRATH, J. E., ed. (1970) *Social and Psychological Factors in Stress*, Holt, Rinehart & Winston.

MCKINSEY ASSOCIATES (1968) 'Unlocking the computer's profit potential', *The McKinsey Quarterly*, **5**, no. 2.

MANN, F. C. and NEFF, F. W. (1961) *Managing Major Change in Organizations*, Ann Arbor, Michigan, Foundation for Research on Human Behavior.

MARCH, J. G., ed. (1965) *Handbook of Organizations*, Rand McNally.

MARCH, J. G. and SIMON, H. A. (1958) *Organizations*, Wiley.

MAUNDER, M. and STYLES, A. (1969) 'A formal training scheme for systems analysts', *Computer Weekly*, 24 April.

MECHANIC, D. (1962) 'Sources of power of lower participants in complex organisations', *Admin.Sci.Q.*, **7**, no. 3, 349–64.

MELMAN, S. (1958) *Decision Making and Productivity*, Wiley.

MERTON, R. K. (1949) *Social Theory and Social Structure*, Free Press of Glencoe.

MORRIS, J. (1971) 'Management development and development management', *Personnel Review*, **1**, no. 1, 30–43.

MORRIS, W. T. (1964) *The Analysis of Management Decisions*, Irwin.

MUMFORD, ENID (1967) 'A survey of management services staff in a large retailing organisation', unpublished paper, Manchester Business School.

MUMFORD, ENID (1968) 'Planning for computers', *Management Decision*, Summer, pp. 98–102.

MUMFORD, ENID (1972) *Job Satisfaction: a study of computer specialists*, Longman.

MUMFORD, E. and BANKS, O. (1967) *The Computer and the Clerk*, Routledge & Kegan Paul.

MUMFORD, E. and WARD, T. B. (1966) 'Computer technologists: dilemmas of a new role', *Journal of Management Studies*, **3**, 244–55.

MUMFORD, E. and WARD, T. B. (1968) *Computers: planning for people*, Batsford.

NADEL, S. F. (1957) *The Theory of Social Structure*, Free Press of Glencoe.

NEWMAN, W. H. (1951) *Administrative Action: the techniques of organization and management*, Prentice-Hall.

OZBEKHAN, H. (1969) 'Toward a general theory of planning', in E. Jantsch, ed., *Perspectives of Planning*, Paris, OECD.

PALME, J. (1973) 'Computers in the 1980's', *Management Informatics*, **2**, no. 4, 173–6.

PETTIGREW, A. M. (1968) 'Inter-group conflict and role strain', *Journal of Management Studies*, **5**, no. 2, 205–18.

PETTIGREW, A. M. (1972a) 'Managing under stress', *Management Today*, April.

PETTIGREW, A. M. (1972b) 'Information control as a power resource', *Sociology*, **6**, no. 2, 187–204.

PETTIGREW, A. M. (1973a) 'Occupational specialisation as an emergent process', *The Sociological Review*, **21**, no. 2, 255–78.

PETTIGREW, A. M. (1973b) *The Politics of Organisational Decision Making*, Tavistock Publications.

PONDY, L. R. (1967) 'Organisational conflict: concepts and models', *Admin.Sci.Q.*, **12**, no. 2, 269–320.

RAIFFA, H. (1968) *Decision Analysis: introductory lectures on choice under uncertainty*, Addison-Wesley.

RANDALL, R. (1973) 'Influence of environmental support and policy space on organizational behaviour', *Admin.Sci.Q.*, **18**, no. 2, 236–47.

RAPPAPORT, A. (1967) 'Sensitivity analysis in decision-making', *Accounting Review*, July, pp. 441–56.

REDDINGTON, M. (1972) Unpublished survey, National Computing Centre.

REID, J. G. (1968) 'Management services – innovators or helpers?', *Work Study and Management Sciences*, December, pp. 732–8.

REX, J. (1961) *Key Problems of Sociological Theory*, Routledge & Kegan Paul.

RITTEL, H. W. J. and WEBBER, M. M. (1973) 'Dilemmas in a general theory of planning', *Policy Sciences*, **4**, 155–69.

ROGERS, E. M. (1962) *Diffusion of Innovations*, Free Press of Glencoe.

RUBINSTEIN, A. *et al.* (1967) 'Some organizational factors related to the effectiveness of management science groups in industry', *Management Science*, **13**, no. 8, 508–518.

SACKMAN, H. (1971) *Mass Information Utilities and Social Excellence*, New York, Auerback.

SHAFER, R. J. *et al.* (1967) 'What is national planning?', in Gross (1967).

SCHWARTZ, M. H. (1969) 'Computer project selection in the business enterprise, *Journal of Accountancy*, April.

SCOTT, W. H., MUMFORD, E., MCIVERING, I. and KIRBY, J. (1963) *Coal and Conflict*, Liverpool University Press.

SHERIF, M. (1963) *The Psychology of Social Norms*, Harper & Row.

SIMON, H. A. (1956) 'Rational choice and the structure of the environment', *Psychological Review*, **63**, 129–38.

SIMON, H. A. (1961) *Administrative Behaviour*, New York, Macmillan.

SIMON, H. A. (1965) 'The new science of management decision', in *The Shape of Automation for Men and Management*, Harper & Row.

SOELBERG, P. (1963) 'The structure of individual goals: implications for organisational theory', in G. Fisk, ed., *The Psychology of Management Decision*, CWK Gleerup Publishers, Lund Sweden, 1967.

STRAUSS, A. *et al.* (1963) 'The hospital and its negotiated order', in E. Friedson, ed., *The Hospital in Modern Society*, Free Press of Glencoe.

STRAUSS, A., SCHATZMAN, L., BUTCHER, R., EHRLICH, D. and SABSHIN, M. (1964) *Psychiatric Ideologies and Institutions*, Free Press of Glencoe.

TAYLOR, B. and IRVING, P. (1971) 'Organized planning in major U.K. companies', *Long Range Planning*, June.

TAYLOR, D. W. (1965) 'Decision-making and problem-solving', in March (1965).

THOMPSON, J. (1967) *Organizations in Action*, McGraw-Hill.

TRIANDIS, H. C. (1960a) 'Cognitive similarity and communication in a dyad', *Hum. Rel.*, **13**, 175–83.

TRIANDIS, H. C. (1960b) 'Some determinants of interpersonal communication', *Hum. Rel.*, **13**, 279–87.

TRIANDIS, H. C. (1967) 'Interpersonal relations in international organisations', *Organisational Behaviour and Human Performance*, **2**, 26–55.

TURNER, R. H. (1962) 'Role-taking: process versus conformity', in A. M. Rose, ed., *Human Behaviour and Social Processes*, Routledge & Kegan Paul.

WALKER, K. F. (1971) 'Personnel and social planning on the plant level', in G. Friedrichs, ed., *Computer und angestellte*, Frankfurt, Europäische Verlagsatstalt.

WALTON, R. E. (1965) 'Theory of conflict in lateral organizational relationships', in J. Lawrence, ed., *Operations Research and Social Sciences*, Tavistock Publications, 1965.

WEINSTEIN, K. (1971) 'Computer specialists: a study in role ambiguity', unpublished MSc thesis, University of Manchester, Institute of Science and Technology.

WHYTE, W. F. (1961) *Men at Work*, Irwin, Dorsey.

WIEST, J. D. (1967) 'A heuristic model for scheduling large projects with limited resources', *Management Science*, **13**, no. 6, 8359–77.

WILDAVSKY, A. (1964) *The Politics of the Budgetary Process*, Little, Brown.

WILENSKY, H. L. (1967) *Organizational Intelligence*, New York, Basic Books.

WILLIAMS, C. W. (1971) 'Allocations for the development and employment of scientific resources: a conceptual framework', in M. D. Rubin, ed., *Man in Systems*, Gordon & Breach.

WOODWARD, J. (1965) *Industrial Organization: theory and practice*, Oxford University Press.

Index